THE

LOUISIANA SEAFOOD BIBLE

THE
LOUISIANA
SEAFOOD
BIBLE

Fish Volume 1

Jerald and Glenda Horst

PELICAN PUBLISHING COMPANY
GRETNA 2014

First printing, September 2012
Second printing, September 2014

Library of Congress Cataloging-in-Publication Data

Horst, Jerald.
  The Louisiana seafood bible. Fish / Jerald and Glenda Horst.
     p. cm.
  Includes index.
  ISBN 978-1-4556-1527-8 (hardcover : alk. paper) — ISBN 978-1-4556-1528-5 (e-book)
 1. Cooking (Fish) 2. Cooking, American—Louisiana style. 3. Fish trade—Louisiana. I. Horst, Glenda. II. Title.
  TX747.H673 2012
  641.6'92—dc23
                                        2012014741

Printed in China

Published by Pelican Publishing Company, Inc.
1000 Burmaster Street, Gretna, Louisiana 70053

The sea hath fish for every man.

William Camden (1551-1623),
English antiquarian, historian, topographer, and officer at arms

They say a fish should swim thrice. . . . First, it should swim in the sea, then it should swim in butter, and at last, sirrah, it should swim in good claret.

Jonathan Swift (1667-1745),
Irish cleric, satirist, political pamphleteer, and author

Soup and fish explain half the emotions of human life.

Sydney Smith (1771-1845),
English cleric and writer

Teach a man to fish and you feed him for a lifetime. Unless he doesn't like sushi—then you also have to teach him to cook.

Auren Hoffman, (1974-),
Internet Herald editor and philosopher

Contents

Part II: Recipes

Preface

This, the fifth volume of the *Louisiana Seafood Bible,* is the first of the final two volumes in the series. Both volumes are devoted to fish, the kind with fins, often referred to by biologists as "finfish." The first four volumes in the series were given over to shellfish: shrimp, crawfish, crabs and oysters. Over 120 species of edible fish are taken from Louisiana and its nearby Gulf of Mexico waters, far too many to comfortably discuss in one volume.

The information in this book is the fruit of a thirty-year career with Louisiana State University, in which I worked with the state's seafood industry, and of Glenda's background growing up as the daughter of a commercial fisherman in the small fishing community of Bayou Sorrell. A major portion of my responsibility at LSU was to develop and transfer to the fishing industry better methods of harvesting, processing, and marketing Louisiana's abundant fisheries resources.

As is usually the case, information flowed both ways. I learned much about the intricacies of the world of fisheries from the people who made bringing seafood to peoples' tables their lives, as well as livelihoods.

During the course of my career, I published a monthly fisheries newsletter called *Lagniappe.* Each edition ended with a recipe. Some of those recipes came from Glenda and my personal test kitchen. We love to cook and we love seafood.

A larger number of recipes were sent in voluntarily by the readers of the newsletter, with the understanding that they would be publically shared. Glenda and I tested each and every recipe we received, and we tested them against each other. The recipes printed were the best of the best. Many of the recipes in this book were originally published in that newsletter.

Glenda and I also love to travel and eat. We have been fortunate enough to meet many others who love seafood as we do. Many of these people have worked within or have connections to the seafood industry, but a surprising number were dedicated seafood lovers from all walks of life. These people invited Glenda and me into their homes to share the preparation and enjoyment of their prized family recipes. These experiences have been beyond delightful and have left us with many warm memories, as well as world-class recipes.

This book is a joint effort with my wife and cooking partner, Glenda. It is written in the first person by me only for the sake of being easier to read than if the reader had to switch between the two of us constantly.

Get hooked on fish!

Jerald Horst

Acknowledgments

Many obligations of thanks are owed to the people who helped make this book possible. We are grateful to all whose help was so generously offered. Jane Black, Craig Borges, Benny Champlin, John Dupuis, Pete Gerica, David Guilbeau, Patti Hulett, Randy Montegut, Harlon Pearce, Robert Romaire, and Robert Wegmann Jr. answered many questions for us and helped us clear many hurdles in our field work. Thanks are extended to Pat Attaway for modeling the fish filleting process.

We received a great deal of assistance with access to historic and modern photographs thanks to the help of Benny Champlin, Robert Fritchey, Mark Shexnayder, Lawrence "Brother" Stipelcovich, and Sharron Terry. Sincere thanks are extended to the following institutions and individuals for generously sharing their photographs and images: Mary Abadie, Gerald Adkins, Daniel Alario Sr., Preston Battistella, Alana Jo Beaugez, Tommy Cobb, Byron Despaux, the Historic New Orleans Collection, the *Jefferson Parish Yearly Review,* William Knipmeyer, the Louisiana State Archives, the Louisiana State Museum, the Louisiana State University AgCenter, Greg Lutz, and the Morgan City Archives.

We are indebted to Larry Kenneth "Chico" Moore for welcoming us onto his fishing boat with our intrusive camera and sharing his world with us. We also want to thank the people who have donated recipes to the *Lagniappe* newsletter through the years.

Deepest thanks are extended to the following individuals who welcomed us into their homes and businesses while they prepared the recipes that they shared with us: Tracey "T. L." Bayles, David Bertrand, David "T-Coon" Billeaud, Deano and Jacquie Bonano, Benny Champlin, Melanie Charpentier, Reggie Craig, Louis "Woody" Crews, Aulton Cryer Jr., Gayle Daigle, John Davis, John Dupuis, John Falterman Jr., Gary Hayes, Bert Istre, Glinda Jenkins, Lane Lemaire, Jason Lobello, Chris Macaluso, Brandi and Todd Masson, Mike "Road Kill" McMullen, Rusty Munster, Ray "Garfish Ray" Ohler, Jack and Janis Oser, Alex Patout, Barbara Wurzlow Picard, Mary Poe, Anthony Puglia, Robin and René "Stretch" Reid, David Robinson, Larry Roussel, Tina Rue, Bill Spahr, Donald Spahr, Terry "Duke" St. Cyr, Kaleigh Stansel, Lawrence "Brother" Stipelcovich, John Supan, Duane Taylor, Mark Vicknair, and Shane Zeringue.

Part I: Fish

A Wealth of Fishes

Louisiana, a land seemingly half water and half dry ground, is home to a cornucopia of fish, both fresh and saltwater. The native tribal groups the Chitimachan, the Muskhogean, the Attakapan, the Natchezan, the Tunican, and the Caddoan all made use of fisheries resources in their diets. Nets and traps made of degradable materials have disappeared, but some stone fishing points and net weights have been found.

Coastal Indians left a more solid record with the presence of their shell middens. Essentially, the middens were garbage dumps composed largely of clam and oyster shells. Mixed in with the shells were items that would otherwise have been lost in the background soils of the state's wetlands. Included in these were broken pottery, pieces of or whole worked stone points, large and small animal bones, and a few fragile fish bones.

Human consumption of fish is known to have gone far back before the birth of Christ. But the conversion of Europeans to Christianity, a religion with strong ties to fish and fishing, had a profound impact on both the Old and New Worlds. Peter, Andrew, James, and John, the

first four disciples chosen by Jesus to follow him, were commercial fishermen. Mark 1: 16-20 narrates the event.

> As he passed by the Sea of Galilee, he saw Simon and his brother Andrew casting their nets into the sea; they were fishermen. Jesus said to them, "Come after me, and I will make you fishers of men." Then they left their nets and followed him. He walked along a little farther and saw James, the son of Zebedee, and his brother John. They too were in a boat mending their nets. Then he called them. So they left their father Zebedee in the boat along with the hired men and followed him.

The New Testament makes repeated mention of fish and nets, including the story of the large fish that the Lord sent to swallow Jonah, the disobedient prophet. But nothing symbolizes Jesus' ministry more than the miracle of feeding the multitude with five loaves of bread and two fish, as related in Mathew 14: 13-21. Jesus had just heard of the death of John the Baptist and had withdrawn to a remote area near Bethsaida.

Tilapia are believed to almost certainly be the fish harvested from the Sea of Galilee. (Courtesy Greg Lutz)

When Jesus heard of it, he departed thence by ship into a desert place apart: and when the people had heard thereof, they followed him on foot out of the cities. And Jesus went forth, and saw a great multitude, and was moved with compassion toward them, and he healed their sick.

And when it was evening, his disciples came to him, saying, This is a desert place, and the time is now past; send the multitude away, that they may go into the villages, and buy themselves victuals. But Jesus said unto them, They need not depart; give ye them to eat.

And they say unto him, We have here but five loaves, and two fishes. He said, Bring them hither to me. And he commanded the multitude to sit down on the grass, and took the five loaves, and the two fishes, and looking up to heaven, he blessed, and brake, and gave the loaves to his disciples, and the disciples to the multitude.

And they did all eat, and were filled: and they took up of the fragments that remained twelve baskets full. And they that had eaten were about five thousand men, besides women and children.

Not surprisingly, early Christians adopted the fish, ixthus, as their symbol. The connection was permanent. The cross, with its symbol, followed the sword as European explorers colonized the Americas. Early Europeans made frequent note of the number and variety of fish in North America. Louisiana was no exception. The French adventurer, Antoine LePage du Pratz, in his *Histoire de la Louisiane,* the memoirs of his observations in the French colony of Louisiana between 1718 and 1734, describes the fish he found.

Though there is an incredible quantity of fishes in this country, I shall be concise in my account of them; because during my abode in the country

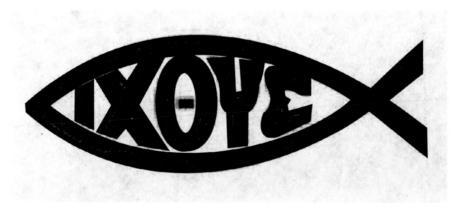

Early Christians adopted the fish as their symbol.

they were not sufficiently known. . . . The Choupic is a beautiful fish; many people mistake it for a trout, as it takes a fly in the same manner. But it is very different from the trout, as it prefers muddy and dead water to a clear stream, and its flesh is so soft that it is only good when fried.

While little written account of Louisiana's fisheries exists from colonial times to the American Civil War, fish and other seafood were certainly used. But its consumption was, by necessity, local and seasonal. Nothing could be preserved. Canning didn't exist; ice didn't exist; mechanical refrigeration didn't exist; and transportation was crude or close to nonexistent.

The first form of fish preservation that allowed shipment was drying. Wooden platforms to dry shrimp were erected in the state's marshes by the mid-1860s. Large numbers of speckled trout were dried on the platforms and shipped out in wooden barrels.

A major step forward came with the development of ice-making plants, with the first one along the coast being built by the New Orleans Refrigeration and Manufacturing Company at the foot of St. Charles Street in 1883. Before this, any ice used had to be imported by vessel from New England. Ice, coupled with rail shipping, allowed fish to be transported across the state and out of state.

Commercial fishing for freshwater fish for shipment grew rapidly

Louisiana shrimp drying platforms in the marsh dried speckled trout as well as shrimp. (Courtesy Tommy Cobb, Blum & Bergeron Inc.)

along the north-south axis from Morgan City to Jonesville. Demand in New Orleans markets for saltwater species increased along with the ability to maintain fish quality with ice.

The July 1911 edition of *The Louisiana Grocer* extolled the virtues of the Crescent City's fish supply.

> Fish is one of the favorite articles of food in this city, and the fact that a large part of the population can eat no other meat on certain days leads to a wonderful trade in fish. The methods of cooking fish in this city make it one of the dishes that strangers rave over and it is always conceded that nowhere else in the country will one find such fish, cooked so delightfully. Red fish, sheepshead, trout, Spanish mackerel, flounders, pompano, croakers, these are to be found on most fish stands; but of these the pompano holds first place, with the trout holding second. "Tenderloin of trout," is one of the chief dishes of the best restaurants and when accompanied by sauce tartare it cannot be excelled.

New Orleans had been famous for its fish for some time. In his 1883 book, *Life on the Mississippi,* Mark Twain called attention to one of them: "The renowned fish called the pompano, delicious as the less criminal forms of sin."

Speckled trout have been the most popular saltwater fish in Louisiana for more than a century and remain so.

In 1916, the Louisiana Grocers Association held its convention in New Orleans, and in announcing it, these food experts used the lure of the city's fish in its restaurants to boost attendance.

There are other things which are cooked in New Orleans restaurants as they are nowhere else in the world. Close at hand are all varieties of sea food; and New Orleans has long been distinguished as the one place in which one could find the finest fish and crabs and shrimps that were to be found anywhere. The fish are in great variety—red snapper, Spanish mackerel, sheephead, flounders, croakers, green trout, speckled trout, pompano; some of them aristocrats of the fish tribe, always desirable, and when cooked by a New Orleans chef, not to be equaled.

New Orleans seafood purveyors continued to have easy access to the great variety of saltwater finfish for decades. However, by 1977, Gerry Waguespack, a Metairie recreational boat dealer, and Al Bankston, a Baton Rouge printer, had organized "Save our Specks," with the goal being to outlaw gill nets, the tool of choice for commercial finfishermen.

Seafood dealers such as Ed Martin Seafood Company in the French Market supplied the city with a wide variety of fish. Owner Ed Martin is on the right in this 1930s photograph. (Courtesy Daniel Alario Collection)

The initiative was assumed by the Louisiana affiliate of the Gulf Coast Conservation Association in 1983. Under their guidance, the Louisiana legislature, with Act 889, took redfish off the commercial market. They followed with Act 815 in 1995, which outlawed the use of saltwater gill nets for all but a small number of mullet and pompano fisherman.

Louisiana's other major finfishery, that for freshwater fish, developed differently. The vast maze of rivers, swamps, and lakes held prodigious quantities of fish. But, prior to access to manufactured ice, the main species taken were catfish, which were hardy and could be kept alive relatively easily. What catfish were shipped out of state went mainly to nearby states.

But ice and the ability to ship fish by rail cars changed that, and by 1931, the state's fishermen sold a recorded 19,213,368 pounds of freshwater fish, far more than any other state. In addition to catfish, huge amounts of buffalo and lesser numbers of gaspergou were taken for sale. Buffalo were shipped long distances, with two major markets being the eastern European and Jewish trade, mainly in New York City, but also in Chicago.

The fishery throve until a perfect storm of events crippled it in the early 1960s. Catfish, the "money fish" for commercial fishermen, were utterly decimated in the Mississippi and Atchafalaya river systems following a massive endrin spill into the Mississippi in Memphis, Tennessee, in 1963.

The endrin spill didn't affect the waters of the Red, Black and Ouachita river systems, but in 1964, another event did. Railroads stopped accepting iced fresh fish for shipment. The trucking industry was not yet well developed and the modern interstate highway system was in its infancy. Access to distant markets shriveled.

Concurrent with these problems, Louisiana's burgeoning oilfields were experiencing severe labor shortages. Multigenerational fishermen left the fishery for the "Oil Patch," most to never return once their families became accustomed to a regular paycheck.

Finally, catfish farming became an economic reality. Arkansas was the first state to produce the fish on a commercial level in 1963, followed by Mississippi, the current leader in production, in 1965. While early production filled the void left in the market by the endrin spill, the industry grew to dominate price and demand for catfish, even when the wild fishery returned to health. The freshwater commercial finfishery, like the saltwater fishery, still exists today, although in a much changed form.

Enormous quantities of freshwater fish were purchased by fish dealers such as Blum and Casso Fisheries in Berwick, Louisiana. (Courtesy Morgan City Archives)

Over 120 species of fresh and saltwater fish are harvested for sale, and recreational fishermen have access to a score or more species of game fish, all of which are excellent food fish. Louisiana, more than any other state, does indeed have a wealth of fishes.

Sea Dogs

I lie quietly in my bunk in the dark and cramped fo'castle of the forty-two-year-old fishing vessel. It's an upper bunk and the deck timbers are less than a foot from my face. Glimmers of light reflected off the sea's waves sneak in around the material obscuring a porthole near my head and play across the badly peeling deck and timbers.

A generation and a half of commercial fishermen have laid their weary bones in this spot before me. Three of them, lightly snoring, share this tiny space (called a "dungeon" by one) with me now.

I can't sleep.

Small sounds intrude. The nylon anchor rope groans in protest as each passing wave forces the boat against its anchor 480 feet below. The wood in the old hull creaks in concert with the ropes.

It's 3:00 P.M. on day four of the trip and we've run out of fish. The captain and mate have decided to rest the crew for the afternoon in preparation for a shift of night fishing. They suspect that their primary quarry, vermilion snappers, or "b-liners" in fishermen's lingo, have changed to a night bite.

The morning's catch had been meager, two seventy-pound baskets of the creatures and one lonesome grouper. I lay there wondering how the rest of the trip would pan out.

Five days earlier, I met the four men who work on the *Nite Owl*, a Golden Meadow, Louisiana-based snapper bandit boat. Bandits are the hydraulic fishing reels that the boat is equipped with. In someone's imagination in the past, the stubby rod on each bandit suggested the arm on a slot machine, a.k.a., a one-armed bandit.

The owner and skipper of the *Nite Owl* is Larry Kenneth Moore, always referred to as "Chico," a name he picked up in construction work years ago. "Back then, I looked like a Mexican, I was so dark," he said with his ever-present grin. "Not ten people in Louisiana know my real name."

Chico is short, a little roly-poly, and wears a distinctly Santa Claus-like beard. He is fifty-five and has been commercial fishing for snappers and groupers, collectively known as reef fish, since 1986.

"A friend and I were going to New Orleans for Mardi Gras that year with 2 fifths of everclear liquor and we wound up in Destin, Florida,"

Capt. Chico Moore, owner and skipper of the Nite Owl.

he explained. "I liked it so much that I moved there to do framing work. I was renting a room from a b-liner fisherman and he talked me into trying fishing. I went out eight days and liked it so much that I came in and gave my notice. I've been fishing ever since."

Another fisherman brought him to Louisiana to fish for b-liners and silkies (queen snappers), and he moved to Golden Meadow in 1988. He captained bandit boats from 1990 to 2003, when he bought the *Nite Owl*, a boat originally built to fish in the North Atlantic.

The *Nite Owl*'s first mate, fifty-one-year-old Charles "Chuck" Patrick, carries the air of someone who doesn't mind assuming responsibility. Dark-haired, powerfully built, and seeming to be taller than he really is, he attacks his work with gusto. A knob of silvery chin whiskers perfects the image of a seaman.

Originally from Bonnie Lake, a town near Seattle, Washington, Chuck was on a drive around the country when he found Louisiana seven years ago. "I really liked it," he said about Louisiana. "I made a trip on a longline grouper boat and I liked it too. Then I found that I like bandit fishing better."

Within weeks after going to work with Chico, Chuck rode out Hurricane Katrina on the *Nite Owl* in Bayou Lafourche. Between 100 and 150 boats sunk in the bayou that day.

First mate Chuck Patrick likes the challenge of commercial fishing.

"I love it," he said with an open candor. "I get paid to fish. It's hard work, but a challenge to see how good you can be—how much torture you can put yourself through when catching fish—how long you can stay on your feet. I really like it on blue water. You can see seventy-five feet down. I see things that most people never see, like whale sharks and giant sea turtles."

The senior member of the crew is John Harris, a slender, well-built sixty-six-year-old, with curly brown hair, a droopy moustache, and strong Anglo-Saxon features. Like Chico and Chuck, John hails from outside Louisiana, a common thing in Louisiana's offshore finfish fleet.

Originally from the Eastern Shore of Maryland, John worked on oyster, clam, and crab boats in the Chesapeake Bay for years before moving to Madeira Beach, Florida, to work on a bottom longline boat. He longlined for grouper, shark, and swordfish before settling in on bandit boats.

Iron-willed, articulate, well-read, and opinionated, John has worked a variety of jobs when times were hard in fishing, including a hitch as a gravedigger. "It wasn't like it is now. It was pick and shovel. Oh, that ground was hard. I had to quit school in the tenth grade

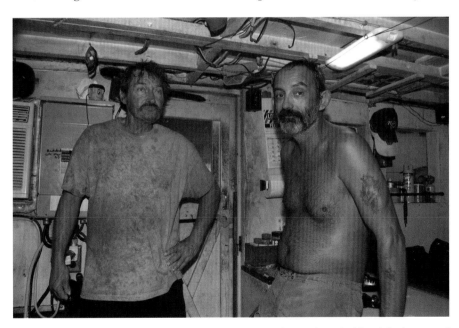

John Harris on the left worked the bandit reels on the starboard side of the boat, and Jerry "T. J." Rogers Jr. cut bait and gutted, washed, and iced the crew's catch.

when Dad died; I was the oldest of eight kids. I have worked hard all my life; never had a desk job. If I retired now, I wouldn't be worth a damn in five years."

John has worked for Chico for a little over a year.

The last member of the crew, Jerry Rogers Jr., was the youngest at forty-four and the only native Louisianian, born and bred in Golden Meadow. T. J., as he is known to everyone, is wiry and spry and readily admits to being "hyperactive." His grizzled beard, dark complexion, and glinting black eyes give him the appearance of a pirate, a term he used to describe himself when he wasn't lamenting being born too late to be a hippie.

Besides being in constant motion, T. J. proved to be the crew's chatterbox, offering constant commentary on whatever was going on in his thick Cajun accent. Much of what he said was in the form of double entendres. Looking at his face didn't help with interpretations. His eyes were usually hidden behind wrap-around sunglasses, and his grin managed to be both wicked and good-natured at the same time.

T. J. has worked on commercial fishing boats off and on since he was sixteen years old and this was to be his second trip with Chico.

The night before the *Nite Owl* sailed, I joined the four men at a local restaurant and watering hole. Over ice-cold Coronas, we talked mostly about the upcoming trip, but I caught Chico eyeing me.

Having my attention, he spoke directly to me in his high-pitched voice. "I need to warn you. We will be over a hundred miles offshore and once we are out there, we don't run in, even if it gets rough." He was smiling, but he was watching my reaction closely.

I wondered what I was getting myself into. At the dock again, I stood back and eyed the *Nite Owl.* The old sharp-prowed corsair was low-slung and looked far too fragile to work 130 miles offshore. But I laid aside emotion, reasoning that anything that could take the other men out and back on repeated trips could safely deliver me to shore on one trip.

Preparations the next day seemed to be endless. Groceries and provisions for ten days had to be purchased and stowed. A mountain of ice was blown into the hold and then bait—frozen mackerel and squid—was purchased and loaded. Finally, the vessel topped off its freshwater tanks and took on fuel.

Everyone was ready to go. While fueling, T. J. bounced and danced around on the back deck plucking on an imaginary guitar, something he did constantly for the whole trip. "I'm ready to go; I'm ready to go," he kept repeating.

The Nite Owl *is built and maintained for seaworthiness, not beauty.*

T. J. danced and strummed on an imaginary guitar the entire trip.

Finally, at 6:00 P.M., the *Nite Owl* left the protected waters of the pass and entered the Gulf of Mexico. In the two- to three-foot seas of the open water, the boat came to life under my feet. At 6 knots, we would have to run all night and part of the morning to get to our first fishing spot.

In the remaining daylight, the three crewmen readied the boat and their fishing equipment for the next day while I chatted with Chico about his plans for the trip. If all goes the way he would like, he will stop in what he calls "shallow water," two hundred or so feet deep, to fish for red snappers the first day. He hopes to catch enough of these, the highest-priced fish that he will catch, to pay the expenses for the trip before running for deeper water.

In deeper water he will target b-liners (vermilion snappers), the bread and butter of the trip, with a couple of forays into deep-water grouper habitat. His fishing strategy is explained by the complicated regulations that govern fishing for snappers and groupers, collectively known as reef fish.

A commercial reef fish fisherman must possess federal fishing permits, the most important of which is the reef fish permit. New reef fish permits are no longer issued by the federal management agency,

Red snappers are the most highly regulated and the highest-valued of the Gulf of Mexico's reef fish species.

Chico hopes to catch enough red snappers and scamp groupers early in the trip to pay for expenses.

the National Oceanic and Atmospheric Administration. A fisherman seeking to enter the fishery must purchase or otherwise obtain a permit being held by another fisherman.

Other permits that Chico holds are a deep-water grouper endorsement, a shallow-water grouper endorsement, a red snapper endorsement, and a gag grouper endorsement. But permits alone are not enough.

A fisherman must also own (or lease from other fishermen) an annual quota amount under each endorsement. These individual fishing quotas are measured in pounds, and once a fisherman has landed his quota amount for the year, he can no longer sell those species.

This largely explains Chico's strategy for the trip. The fish with the most market value are red snappers, scamp (a shallow-water grouper), and yellowedge grouper (a deep-water grouper). If he concentrates his fishing on these species, he will exhaust his quotas early in the year, leaving him nothing for later.

B-liners are a reef fish species that are very common and have no individual quota, so when possible, Chico tries to make them the bulk of his catch. Red snappers and any scamp that he can catch early in the trip are to pay the expenses of the trip. High-value yellowedge grouper catches made later in the trip would be icing on the cake.

My first night out, I melted into a deep sleep, lulled by the steady rumble of the 871 Detroit Diesel and rocked by the gentle sea.

Day 1 at Sea

After riding all night, I crawled out of the dark fo'castle and was greeted by a brilliant sun and azure blue waters. Chico was at the wheel, watching his electronics. A laptop computer with a bathymetric chart on its screen lay on the bunk next to him. A split screen on the device on the console in front of him displayed a depth sounder/fish finder on one side and a GPS screen and plotter on the other half.

Out on deck, radio speakers blared 1960s and 1970s rock and roll from the XM satellite radio over the throbbing sound of the diesel engine. Chuck, John, and T. J. cheerfully prepared for action at our first projected destination, an oil and gas platform in 198 feet of water. T. J. danced and bobbed to the music while he cut Atlantic mackerel into bait-size chunks. Frequently, he interrupted his work on the bait to take a deep drag on a Kool cigarette and use his razor-sharp knife to beat out part of a tune he particularly fancied on his cutting table.

The *Nite Owl's wheelhouse is a maze of electronic equipment jammed into a tiny space, which doubles as Chico's bunk room.*

The twenty-two hooks of a single gear, hanging from the cabin awning for baiting, look like a hopeless tangle but are ready to fish.

Less theatrically, the other two men rigged their gears, their term for the terminal tackle that they attached to the business end of the three-hundred-pound test line on their bandits. Each gear bristled with twenty-two circle hooks, each of which was attached to the main line or "backbone" of the gear with a fourteen-inch long, 125-pound test monofilament line, alternately called a "snood" or a "snoot." A five-pound or larger lead weight was clipped on the bottom of each gear.

Fully extended, a twenty-two-hook gear extends to more than twenty-five feet long. Shorter gears, with fewer hooks, explained Chuck, are used for other species, typically eight hooks for silkies and four hooks for groupers. Chuck was responsible for the two bandits on the port (left) side of the boat while John manned the starboard (right) side reels.

As soon as the *Nite Owl* was tied off to the bright-yellow-legged platform, all four gears were lowered into the water, or "downed" in fishermen's parlance. Immediately, the stout arms that served as rods on the bandits began to bounce vigorously as fish after fish took the bait. The men grinned at each other knowingly. The air was filled with expectation.

John and Chuck decided to bring up the gears of the front pair of bandits and open the hydraulic valves that power them. To their dismay, the gears of the two reels were wrapped and tangled about each other. Chuck brought in the tangled mess, with twenty-one bright orangey-pink red snappers and a dusky almaco jack, over the rail on his side of the boat. T. J. jumped on the mess with his fish de-hooker, while his partners went to the boat's stern to retrieve the lines of the other two bandits.

After twenty more gleaming red snappers were taken off the lines and tossed into plastic baskets, the stern lines were quickly rebaited and downed. Chuck and John patiently untangled the gears of the front two reels, rebaited them, downed them, and hustled to the back of the boat to haul in the other two lines.

John's gear came up with another dozen jewels, but something big was hooked on Chuck's line. The powerful fish surged, pulled his gear under the boat, and wrapped it in the boat's propeller, breaking the gear in two. The fragment he got back held nine red snappers. Tangled gears, I quickly learned, are one of the occupational challenges that bandit fishermen face.

Fearing that the heavy weight attached to the end of Chuck's line will damage the *Nite Owl*'s hull when the propeller is engaged, the men decided to try to retrieve it. After the front two bandit's gears are brought up, with only a few fish each, the two men used a stout line to sweep the bottom of the vessel from stem to stern. Three sweeps yielded nothing. Apparently, the weighted line dropped off the propeller when the fish broke the line.

Prudently, the men decided to down only two gears at a time, Chuck's front one and John's stern one. But the action was over. After a long soak, one reel held four red snappers and the other produced five. Chico quickly decided to move to another location.

En route, he explained, "We might have lost the fish because we didn't have bait in the water for a while. Or it could have been big fish—sharks, groupers, or big jacks—that moved in and caused the snappers to quit. Or it could be that someone else just fished the spot."

I stood in the cabin and listened. There was no place to sit down except for the captain's chair. This was a working boat.

During the hour and fifteen minute run to the next platform, Chuck and John took a nap. Deck hands on commercial fishing boats of all types will usually grab sleep during any break. They never know when they will have to fish twenty-four hours or more non-stop. Ever-active T. J. was too wired up to sleep.

As the *Nite Owl* ground up to the next platform located in 236 feet of water, Chuck clambered up on the bow to tie the boat off to it, a job made easier by the steadily calming seas.

It was even slower here, with catches averaging four or five red snapper per down. Both John and Chuck vigorously yanked their

Snowy grouper are one of the less common but beautiful grouper species caught.

lines sideways to bounce their gears and attract fish but to no avail. Chico stepped out of the cabin to inspect the catch and his disgust showed. In spite of one down yielding nine snappers on a gear, he decided to leave.

"Six miles," he announced as the distance to the next rig. We were steadily fishing our way out to deeper water where b-liners live. "Last trip we got $4.60 to $4.85 for our [red] snappers. We need to get five hundred pounds of them before we get to b-liner bottom."

This platform was located in 254 feet of water. It was packed with fish. Every down, ten to sixteen fish would come up on a twenty-two-hook gear, including a couple of grouper. The men like groupers. John hefted a ten-pound scamp grouper and looked at it admiringly. "Oh, that's a nice fish. Damn nice! That's a money fish. Um, um; that's a good fish."

Some of the red snappers were big, more than twenty pounds. Then the groupers really started coming. Every down yielded three or four, all scamp except for one thirty-pound or so warsaw grouper. When T. J. gutted the first grouper, he carved out its heart, popped it

T. J. downed the raw heart of one of the first grouper of the trip.

into his mouth raw, chewed it, and swallowed it. "It's a tradition," he said impishly, waiting for my reaction.

The long gears of red and vermilion snappers coming up in unbelievably blue water looked like strings of giant Christmas lights. The mouths and jaws of the scamp were splashed with bright canary yellow color pigment. It was eye-candy.

The men were in high spirits. The radio was blasting out late '60s rock and roll, but T. J. was too busy to dance to it. He was having trouble keeping up with gutting, washing, and icing the fish as fast as his deck mates were catching them. John muttered to himself constantly. The more fish he caught, the more he muttered.

Finally, the bite slowed down for groupers. Although they were still catching red snappers, Chico decided to pull off the fish. He didn't want to exhaust his limited annual red snapper quota too early in the year. He turned the boat south for the three-hour run.

The destination was the 52-Point Lump. The term "lumps," as used by Louisiana fishermen, refer to rocky steep-sided but flat-topped miniature mountains that are strung like beads on a necklace paralleling the Louisiana coast in waters 600 to 1,200 feet deep. Called "diapers" by oceanographers, these features are created by extrusions

Gears loaded with brightly colored fish spiraling out of the blue depths were a beautiful sight.

from deep salt layers that have pierced through the overlying layers of rock and mud. Many are encrusted by soft and even hard corals and are ideal habitat for reef fish such as snappers and groupers.

The 52-Point Lump rises from more than 600 feet of water to a depth of 361 feet. Chico cruised over the lump, watching his chromoscope carefully to mark signs of fish. By 7:00 P.M., the *Nite Owl* was riding on anchor and the men had baited gears heading toward the bottom. Supper that night was "chicken sumpin," a sauced chicken dish prepared during the ride to the lump by Chuck, who doubled as the crew's cook.

The b-liner bite was slow; four to eight fish per down. John gave me a tutorial on the fish. "You have to be patient with b-liners to get them to bite. They are not like red snappers. If they are hungry, they will just go grab the bait, but if not . . ."

"Yeah, b-liners are finicky," interjected T. J.

Chuck finished John's explanation, since John was busy unhooking fish from his gear. "Once you get them biting, you got to keep them going. You have to keep gear in the water."

As soon as one man brought the gear of one reel on his side up, he dropped the other reel's pre-baited gear to the bottom. Baited gear was always down, except when the inevitable tangles between reels occurred.

B-liners are perhaps even more beautiful fish than red snappers. Living up to their official common name, vermilion snapper, their general body color is bright red, much more so than that of their more famous cousin, the red snapper. Narrow yellow lines trace horizontally along the lower flanks of the fish. On the upper half of each fish, much smaller broken blue lines angle upward and rearward from the lateral line.

To top off their gaudy coloration, the outer edge of the dorsal fin on the back of each streamlined fish is trimmed with a broad bright yellow strip along its entire length.

But they are small creatures, averaging maybe one or two pounds. A few push three pounds and many are half-pound fish. It took a lot of them to fill a seventy-pound basket, especially with catches being so small. With the slow bite, it seemed to take forever to fill a basket.

The men kept fishing in the soft dark night air. Horizon to horizon, no lights were visible in any direction, except for one solitary oil platform in the distance. Flat blackness surrounded the boat floating on the inky water. Nothing existed to indicate that it floated over the small mountain beneath it.

B-liners are small but beautiful fish.

The first day, the crew fished through sundown and into the night.

At 11:30 P.M., the men gave up after boating a meager three hundred pounds of b-liners for their effort. Since the men had been fishing right in the middle of shipping lanes, the highways for offshore cargo vessels, Chico picked up anchor and moved the boat outside of the lines to anchor in relative safety for the night.

Day 2 at Sea

The crew of the *Nite Owl* was awake by 7:00 A.M. and moving the boat back to the lump to fish for b-liners. Once on the lump, Chico carefully positioned the boat so that when it drifted back on its anchor rope it would be directly over the fish 350 feet down that he had marked on his chromoscope.

The first down produced only a single Creole fish, a ridiculous-looking but beautiful bigeye and a bar jack, but no b-liners. But the second down came up with ten beautiful b-liners. Chico explained that the boat hadn't completely drifted into position yet on the first down. "A hundred feet makes a big difference."

The men began whaling away at the fish. Every down produced ten to fourteen medium to large b-liners. Someone was always hand-over-handing a gear string of flopping rubies over the rail. T. J. could barely gut the fish as fast as the two rail men caught them.

I watched them work in the rolling seas. The three men were so different. Chuck was a big man—solid and immovable as the deck bounced beneath his feet. He was the quietest of the three.

T. J. moved in response to the sea's rhythm as if he had swivels in his knees, hips, and lower back. When his body stopped moving, he would stand with his feet spread wider apart than his shoulders, but even then his head kept moving as if his neck had a spring in it. When he became excited, his eyes bulged expressively.

John moved more slowly and deliberately than the other two, mostly to protect his sixty-six-year-old body during the vessel's ceaseless movements. But he was never left behind by the two younger men. He talked to himself constantly as he fished and a large jack on his line, particularly one that tangled the gears of two reels, would send him into producing a paroxysm of oaths.

The fish were biting with a frenzy. B-liner bites were different from the red snapper bites of the day before. When red snappers hammered the baits, the arms on the bandit would violently throb downward. The more delicate bite of b-liners made the arms twitch and jerk rather than thump.

Baskets are quickly filled with gutted and individually washed b-liners during a good bite.

Shortly after midday a summer rain squall passed through and the fish completely quit biting. Chico speculated that the shift in the wind direction associated with the squall moved the boat off position. The hydraulic winch hauled up the anchor and he moved the boat no more than 250 feet and dropped the anchor again.

Immediately, they began catching fish again, this time a few scamp grouper came up with the b-liners and caused big smiles to appear on the men's faces. They are worth almost twice as much as b-liners, and groupers are everyone's favorite table fish.

It seemed almost too easy. By 8:45 P.M., the men had boated, gutted, and iced 1,050 pounds of fish, and Chico again moved the boat out of shipping lanes to his previous night's anchorage.

I took a delicious shower from the boat's precious freshwater supply, my first in three days, and slept like a baby, rocked to sleep by the gentle seas.

Day 3 at Sea

At 6:30 A.M., the anchor came up and we headed back to the lump to fish b-liners near the same spot as the day before. Everyone was cheerful after the previous evening's showers and the day's good catches.

A lot goes into positioning a boat properly on marked fish. Before Chico can drop anchor, he must mentally calculate how far ahead of the fish to anchor so that the boat can drift back over them. He must also factor in the effects of wind and water currents on the boat's likely position. Then to complicate things, deep currents can sweep the baits off the marked fish even if the boat is properly positioned.

Chico struggled. He picked up anchor and moved repeatedly as down after down produced mostly bonita (little tunny), gear-tangling jacks, and even blackfin tuna.

The weather didn't help. The skies were grim with squalls closing in from three sides. The sea was uneasy and choppy and the water wasn't blue anymore, but rather a leaden gray.

Finally, the boat was buried in a sheet of rain. "This isn't a good day to be anchored in a shipping lane," Chico muttered to himself. Visibility was only a couple of hundred yards. An ocean-going cargo ship could be on top of them in the blink of an eye. "Even with radar, I would rather not be here," he confided.

So up the anchor came again. Chico turned the boat south-southwest toward another lump out of the shipping lanes called the "Whale's Tail." On one edge of the lump, the chromoscope marked a school of some kind of fish with a big red blob trimmed with yellow. "If I run over this without trying it, I'll be up at night," he said as he dropped anchor.

Nothing—no bites! And the weather was still ugly. Everything above the water line was oppressively gray, lit up by an occasional bolt of lightning.

In a steady drizzle, we moved again, to a small rise in 618 feet of water. It was 12:30 P.M. and few fish were in the boat. Here down after down produced small b-liners and mixed small and large queen snappers, fish that the men always referred to as "silkies."

Queen snappers might be the prettiest fish swimming in the Gulf of Mexico. The most brilliant red possible overlays an almost iridescent mother of pearl base color. Their outlandishly large eyes are opalescent red. Unlike other snappers, which tend to be chunky-bodied, silkies are so elongated that their shape gives rise to their other nickname, "ballbat." Their tail is deeply forked, and in large individuals, the upper lobe of the tail sports a streamer well over a foot long.

Chico twice repositioned the boat on the rise. Mercifully, by 3:00 P.M., the drizzle stopped, and by 4:00 P.M., the skies were clearing.

Queen snappers, dubbed "silkies" by the crew, take the prize as the most beautiful fish in the snapper clan.

Although some yellowedge groupers came over the rail mixed with the silkies and b-liners, the amount of fish being caught was still disappointing.

The *Nite Owl* was moved again, to another small rise in 638 feet of water. We were 127 miles offshore, due south of Cameron, Louisiana. "This is as far offshore as I expect to go," explained Chico as he lowered the anchor to the bottom. There were no other boats in sight—no platforms, no place to hide. "It's just you and the Gulf," he added while looking at me over the rims of his glasses.

Here the men made fair, but not spectacular, catches of silkies, some of them pushing fifteen pounds. Mixed with the showy fish were stumpy-bodied, dull, brownish-gray barrelfish up to thirty pounds in weight. While ugly, this member of the butterfish family is so accepted by fishermen as a delectable food fish that it is usually honored with the name "barrel grouper."

After a long day of poor catches, the crew gave up at 8:30 P.M. While Chico steamed north, they silently munched on slices of a ham that Chuck had boiled in the galley earlier. Chico stared at his chromoscope as if transfixed as he checked out area after area. He was determined to mark fish for the next day's fishing before anchoring.

At the Nite Owl's *farthest penetration into the Gulf, they add a few homely, but delicious, barrel grouper to their catch.*

I drifted off to sleep in my bunk to the throbbing of the *Nite Owl*'s diesel engine.

Day 4 at Sea

A little after 6:00 A.M., I woke to thunder and the sound of rain drumming on the deck inches over my head. Chico had hunted for fish until 11:30 P.M., but he hadn't found anything worth trying to fish. So by 8:30 A.M., the boat was anchored back on the same lump as on the second day.

Immediately, the rail men began beating up on big b-liners. Then the wind slowed down and slightly changed direction, swinging the boat off the fish. Chico repositioned the boat. But a school of big, pugilistic amberjacks had moved in and jumped on the hooks as soon as they got into the fishing zone. The hard-fighting fish tangled the reels' gears and knocked the tender-mouthed b-liners off their hooks.

Chuck stoically baited and rebaited lines. But John clearly hated jacks with a profound thoroughness. His denunciations of the creatures were so articulate and profane that they were humorous. To add to John's problems, he awoke in the morning with a severely swollen jaw from an abscessed molar. His jowly appearance led to T. J. irreverently pronouncing him to be "Marlon Brando."

Throughout the battle of the jacks, T. J. was his usual animated self—a perpetual motion machine that couldn't stand still. He constantly danced to the rock and roll music streaming from radio speakers on the back deck. Between cuts, he tapped out tunes with his knife on the cutting table, and periodically, he laid his knife down to pick up an imaginary rock guitar and strum along to some tune.

And he never stopped talking, jabbering in a Cajun-inflected patois on every subject imaginable—from Hitler to Cajun cooking.

The fantastic bite of the early morning had evaporated. Chico moved the boat over and over. He and Chuck worked together with the GPS and the chromoscope to find fish. Finally, they seemed to find fish, but after two downs, the fish disappeared.

Baffled, the two men decided that the b-liners may have shifted to a night-bite and shut down fishing at 12:30 P.M. I followed the rest of the crew as they headed to their bunks and tried to sleep. But sleep wouldn't come. Would this trip be a failure? I wondered as I lay there.

Chuck is the first crew member out of his bunk. He wants to get a jump on cooking supper, poached barrel grouper with a cream cheese sauce. The wind has picked up in the afternoon, whipping the waves

Big gear-tangling amberjacks seemed to take special delight in biting on John's lines.

into white caps. The sky shows unsettled weather, confirming the NOAA Weather Radio's prediction of a low-pressure system moving eastward from waters off the Texas coast.

By 7:00 P.M., the wind is really honking, but the crew is catching lots of medium and large b-liners. Then, at 8:20 P.M., the pitching boat drags its anchor and moves off the fish. The weather radio says the winds are 10 to 15 knots, but they feel like 20 to 25.

A big orange near-full moon rises while Chico struggles to rehook the anchor on the bottom. John and T. J. perch in corners of the cabin like vultures waiting to pounce. They are ready to fish. Chuck keeps himself busy making minor boat repairs. Four times the anchor fails to grab. Then on the fifth try, at 9:20 P.M., it sets.

Out on deck, while the rail men lower their gears, Chuck, who has studied the chromoscope, proclaims hopefully, "There are fish all over this. But I am sure that he [Chico] is nowhere close to where he wanted to be."

The catch isn't spectacular. Small nuisance tinker mackerel are beating the slower-biting b-liners to the hook. The baskets come slow, but they do come. It looks like it will be a grind.

By midnight, the normally ebullient T. J. becomes quiet. Then,

Endlessly, cutting bait for the twenty-two-hook gears of the four bandits is tiresome when mainly unmarketable fish are taking the baits.

another big amberjack grabs a hook on John's gear and completely trashes it, violently upsetting him. "This would drive anyone to drinking. If the pope was out here, he'd drink a barrel of whiskey." He keeps mumbling as he strings a new gear on the end of his line.

Slowly but surely, b-liner catches improve as the number of hyperactive tinker mackerel declines. By 2:00 A.M., few mackerel were being caught and each down was producing nine to fourteen b-liners. The men settle into a steady pace with no breaks. Bait is cut, gears come up, fish are unhooked and hooks are baited, and the gears go back down. T. J.'s knife seems to fly, and when he isn't gutting fish, he is icing them by the basket.

We are now into five days at sea. We are in our own little cocoon. What the stock market is doing is irrelevant. What the president said today doesn't matter. What happened in the Middle East is not on our mind. None of it exists. We are in our own little fifty-foot-by-sixteen-foot world—an island of light and XM classic rock in the darkness, interrupted only by a full moon.

At 4:00 A.M. the b-liners are still biting. Chuck shows a little fatigue when he takes a moment to rest his head on his arm against the pipe-awning frame. T. J. has caught his second wind. John is steadily

Fish are carefully iced to maintain top quality. They are placed one-layer deep on an ice bed in a bin, and then carefully covered with another layer of ice for the next layer of fish.

plugging along determinedly, although his eyes are nearly as puffy as his swollen jaw.

An hour later, the bite begins to flag with two to four b-liners coming per down. The boat has again swung off its spot, but with the approach of daylight, Chico deems it not worth re-anchoring. He pilots the boat out of shipping lanes to anchor in relative safety. The crew turned in to their bunks without showering to save fresh water. They have iced a little more than a thousand pounds of fish.

Day 5 at Sea

The building seas waken the groggy crew at 11:30 A.M. Four- to six-foot seas had built up during the morning and white caps surround the boat. Chico expresses unhappiness that the prediction is for stronger winds and higher seas tomorrow.

It is 2:30 P.M. by the time the boat is anchored securely in 351 feet of water back on the 52-Point Lump and the first gears go down. The seas have built even higher. On the heavily pitching deck, the men struggle to fish, alternately trying b-liner gears and spreader bars, single-hook rigs on long leaders.

The first three drops yield a moray eel, a jack, and a scamp grouper. The men have to fish one-armed, using their other arm to hang onto the vessel as it rises on walls of blue-green water. Subsequent drops yield nothing but jacks and an occasional bonita.

T. J., with nothing to gut and ice, is the first to retreat from the deck to his bunk. John hangs in a little while longer, and then follows his crewmate. Chuck takes a break from fishing but decides to stay up to be ready for the expected night bite. At dusk, he downs his b-liner gear. It comes up with eight respectable fish. At 7:20 P.M., he downs his gear again and goes below decks to tell John and T.J. that the b-liners have started biting. The two quickly jump up and join him on deck.

John lurches onto the deck, completely clad in his foul-weather slicker gear. He eyes Chuck balefully and then he declares, "This is gonna be a miserable night." T. J. is silent.

In spite of the conditions, they do catch fish. For several hours, each down produces eight to ten b-liners, then catches peter out to three to five per down. In the violent weather, Chico hauls in the anchor, repositions the boat, and manages to get the anchor to grab bottom.

The move pays off with eight to twelve b-liners coming up on every down. The rough sea has the men staggering like drunks. The

When the b-liner bite shifts from day to night, the men follow them.

moon is completely buried behind clouds so the men can't see the approaching waves until a wall of black water appears in the halo of their deck lights. The anchored boat climbs each wave, then plunges precipitously down its backside. The backwash from the wave cascades over its stern. Fortunately, the boat is self-bailing and the brine exits on its own, but the decks are always sloshing water.

The three work like automatons, but it is clear that the rough seas are beating them up. Some of their camaraderie is beginning to show signs of wear. We are into the second half of the trip, so a certain amount of griping between them is natural.

At 3:00 A.M., the b-liners quit cooperating. Although they have boated only about five hundred pounds of salable fish, Chico moves to safer anchorage rather than repositioning the boat for more fishing. He is all too happy to leave the shipping lanes. Three large cargo ships, lit up like miniature cities, have passed us while fishing this night.

Day 6 at Sea

The crew of the *Nite Owl* doesn't rouse itself until near midday. They are sore and tired from the battering that they received fishing during the night. On deck, they peer out at diminishing seas and blue skies. Evidently, the bad weather expected out of Texas for today had come through a day early, giving us the sloppy conditions of the day and night before.

After pulling some maintenance on the boat's engine, Chico takes it to another nearby lump, the Three Sisters, to try for some groupers during daylight hours before returning to fish for b-liners on the 52-Point Lump. He anchors in 205 feet of water on top of the lump, the shallowest water that they have fished in for days. The men rig their reels with spreader bars and big baits, trying to entice bites from big groupers.

The first stop yields only a turbo-sized jack, John's ever-present nemesis. Chico moves the boat a short distance. T. J., who is manning one of the reels for variety, gets a hard hit and brings up a twenty-pound marbled grouper. It's an odd-looking creature, with a steeply sloping forehead and a brown body covered with small greenish-blue blotches that look like penicillin mold.

But it's a grouper, and that means that it tastes good and sells for a good price, so the men are happy with it. They catch five more just like it and an outsized red snapper of about twenty-five pounds to boot.

Marbled groupers are an oddity, both in their frequency of catch and their appearance.

When they quit biting, John, who is clearly ready for the trip to end soon, croaks, "I guess that we'll be back on them aggravating-ass b-liners again tonight." Chuck grins and shoots me a look, observing, "Every fisherman likes to catch nice grouper."

But it's still too early to move to the b-liner lump, so Chico moves the boat again to a spot on the same lump 236 feet deep. They are still rigged with spreader bars but entice only one scamp and a whole passel of jacks into biting. Changing from spreader bars to b-liner gears results in only three small red snapper and another gang of jacks.

Just after 6:00 P.M., Chuck declares, "Let's head back to the shipping lanes and see if we'll get lucky and get run over." John visibly winces and grimaces. Chuck enters the cabin to tell Chico that they are ready to go.

By the time the *Nite Owl* is anchored up on the lump, the molten orange sun is approaching the watery horizon to the west. Just enough wind stirs to keep us cool and the seas have calmed off to near flat.

The night is gorgeous and everything is right. String after string of b-liners as bright as shimmering rubies are pulled over the rail. No jacks appear and gear tangles are obvious only by their absence. The only bycatch is the occasional valuable and welcome grouper. Not even the usually omnipresent tinker mackerel are around.

As if to make up for the vicious beating that it administered last night, the sea puts on a spectacular show tonight. Beneath the umbra of the *Nite Owl*'s bright deck lights, the crystalline sea water is the deepest shade of indigo imaginable.

And it teems with life. Squid pause in the lights to look for prey, and then jet off after it with deceptive speed. Long slender needlefish tool themselves through the lights as if on a mission. The boat is surrounded by flying fish, both two-winged and four-winged species. They hover in the water like baby blue-colored butterflies, before launching themselves into the air at the first sign of a larger predator fish. The deck of the boat becomes littered with them.

Predator fish abound, especially bonita, dolphin, and the occasional blackfin tuna. Bonitas and tunas strike small fish like silver missiles and are instantly gone. The dolphin are the easiest to observe and the most beautiful. Their phosphorescent green body color appears shimmering blue beneath the water. Their entire tail and the tips of both pectoral fins are bright yellow. The yellow tips of their pectoral fins look like nothing so much as landing lights on an aircraft's wings.

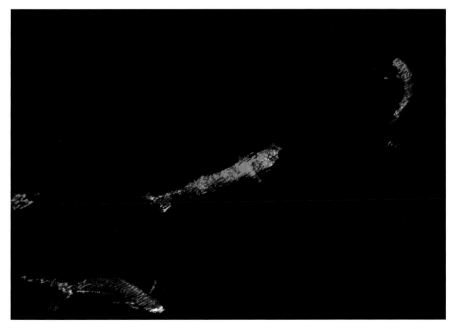

Dolphin light up the inky waters around the boat in a spectacular display the last night of fishing for the men.

Then, at 2:25 A.M., a spectacular sight appears. A large loggerhead sea turtle placidly paddles up directly to the starboard gunnel of the boat, pops its head out of the water, and inspects the men, before submerging and silently slipping away. The three fishermen are left gape-mouthed.

We are floating in a sea of black tranquility.

The men continue to fish until just after 5:30 A.M., when they call it a night. Almost a thousand pounds of fish have been added to the boat's load.

Day 7 at Sea

This looks to be the last fishing day of the trip. Yesterday Chuck asked T. J. if enough ice was left for another night of fishing. T. J. answered in the affirmative, to John's substantial and vocal disgust.

By 1:00 P.M., the *Nite Owl* is heading toward a small lump, where Chico wants to take another stab at groupers and silkies. The day is beautiful. The bright sunlight makes the sea's surface glitter as if it is strewn with millions of diamonds.

The fish prove elusive. The first stop, in 642 feet of water, doesn't produce a bite. Neither does the next stop, at 558 feet. Then in 539

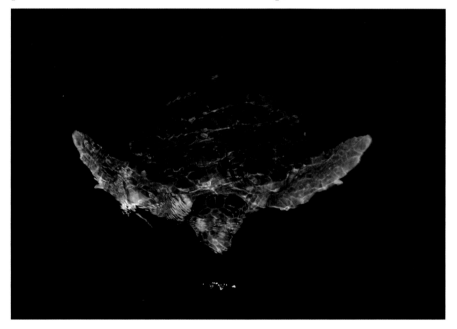

A loggerhead sea turtle caps the night's beautiful sights.

feet of water, they find fish. They pick up eleven nice yellowedge groupers, four snowy groupers, and three big silkies. The catch perks the obviously tired men up a little.

Then, abruptly at 6:00, the bigger fish stop biting and very small b-liners appear. While they are of legal size, no one wants these fish when bigger b-liners are almost certain to be had on the 52-Point Lump.

So following the drill of the last few days, they find themselves anchored on the big lump by 6:45 P.M. They begin catching big b-liners almost immediately. But they are obviously tired and stressed. Chuck handles it by working like a man possessed.

John's growling has become gravelly and persistent. T.J. responds less than good-naturedly. The little things that each man does grates on the others' nerves like fingernails on glass.

But they are professionals and fish until the end they must, so they toil away, mostly now in silence. Catches are good, but it is obvious that the men would have been happy to see the trip end last night.

At the bewitching hour, Chuck leaves the back deck to confer with Chico. He returns to tell John and T. J. to stow their gear and begin clean-up. After the anchor is up, the bow of the boat swings to the north to begin the eighteen-hour ride back to the hill.

The capstone for the trip is bin after bin of well-iced snappers and groupers ready for unloading and shipment to restaurants.

The inside of the cabin is dark. Chico's whiskered face is lit up only by the light emanating from his electronic navigation aids. I stand quietly for a while and then ask him if he ever gets tired of fishing. He answers with a pensive note in his voice. "I still look forward to going out. It's not as much fun as it used to be, but it's still fun.

"Before, you could sell anything that you caught; that was a lot of fun. The rules took a lot of the umph out of it, but there's still nothing I would rather do. I don't like the IFQ [individual fishing quota] system because of the cost of buying quota. I fear that a small number of people will end up with all of it."

He thinks a minute, and then partially reverses himself. "But compared to derby fishing during set seasons, I love it. You don't have to fish in dangerous weather. You don't have to discard fish during closed seasons. And it improves the price because we don't flood the market at one time.

"I knew from the first time I went fishing what I wanted to be, so dreams do come true."

A Life in the Fish Business

For 123 years, Battistella Seafood was a fixture in the business of supplying seafood to New Orleans' markets and famed restaurants. When Preston Battistella, third-generation owner of the business, shuttered its doors in July 2000, by all accounts it marked the end of a fascinating era in the history of seafood in the city.

The earliest record of the business dates to 1877 when Battistella's grandfather, Anthony Battistella operated the business. At some time before Battistella was born in 1925, the business was passed on to Battistella's father, Stanley. At that time, the business was on the corner of Ursalines Avenue and Decatur Street in the city's French Quarter, where it remained until the early 1970s, when Battistella moved it to 910 Touro Street.

Born under the care of a midwife at home at 713 Esplanade Avenue, Battistella was the youngest boy of a family of four boys and four girls. "I was the sugar boy," grinned Battistella. "I could get by with anything."

Anthony Battistella's business at the French Market, circa 1895-1900. (Courtesy Preston Battistella)

Battistella's father's family hailed from northern Italy, near the Swiss border, and had at least one German ancestor. According to Battistella, his father "was nothing but business. His nickname was 'gun' or 'pistol' because he was a quick operator. He never owned an auto; never went on vacation. But he did raise eight kids."

Battistella's mother, Katherine Grecco, was born in Trieste in southern Italy. "My mother was very religious," remembered Battistella. "Dad prayed on his knees every night, but rarely went to church, not like my mother."

Battistella's mother was the family disciplinarian. "The old man didn't spank us," he said, "but he used to raise hell at us. Mom would spank us; she would put a strap to us." He remembers once misbehaving at Catholic school. "A brother sent a note home for me to get signed. I got two spankings that day, one from the brother and one from Mom."

Living on the fringe of the French Quarter, Battistella's father would walk to work every day. The family's kids walked to school at St. Louis Cathedral Catholic School on Dauphine and Dumaine streets. He never remembers anyone being accosted.

"The police patrolled on foot with a stick with a lead tip and a leather strap," explained Battistella. "They would use the stick to tap on the iron curbing to signal the other cops. They knew everybody and everything. They knew who belonged and who didn't."

But the French Quarter did have its seamy side. He remembers at the age of ten, in 1935, Decatur Street as being "bad" with "lots of drunks and working girls." Worse was Gallantine Alley, which he called Murderers' Row, where the French Quarter Flea Market is now. "People were found dead there all the time. They also shanghaied sailors from the area."

But by and large, Battistella called the French Quarter a "neighborhood where everyone knew everyone." He remembers walking to the corner of Canal and Royal streets to watch Mardi Gras parades and roller skating around the French Quarter to shop for his mother.

As a youngster, he got into his share of mischief. His father owned a horse and wagon that he used for seafood deliveries and only for deliveries. But the horse was a blooded race horse, and young Battistella couldn't resist the temptation.

On a Sunday, when the business was closed, the youngster persuaded the wagon driver to saddle the creature for him, and he rode it all day.

The French Quarter of Battistella's youth was a different place than today's Quarter.

Come Monday, the horse was too tired to work. "Dad beat the driver. Things were different in those days. But he never squealed on us."

In 1937, the WPA (Works Progress Administration) renovated the French Market and the senior Battistella moved his retail fish market to Ursulines Avenue and Decatur Street, next to Battistella's Po-boys, his brother Andrew's business, and near the Morning Call Coffee Stand.

Battistella's Uncle Andrew kept milk cans for his restaurant in his father's cooler. The Battistella boys would slip into the cooler when no one was watching and skim the cream off the top of the milk. To replace the volume, the boys added water. "Uncle Andrew had a fit when he found out," grinned Battistella, "but Dad didn't even spank us for that one."

Battistella exerted a strong claim that the name "poor-boy" for the famous New Orleans sandwich was invented by his uncle. Much of the restaurant's business was with Cajun farmers from the countryside who brought their produce to the French Market to sell.

"One day, an old guy who hadn't had much luck selling his produce walked in," said Battistella. "He said, 'I'm hungry; can you give me

Battistella's Seafood Stall in the Vegetable Market, 1936. (Courtesy Collections of the Louisiana State Museum)

something to eat?' Uncle Andrew yelled to the kitchen to make the man a 'poor boy's sandwich,' which he gave him for free. From that, the name caught on."

Friday and Saturday were clean-up days at the French Market. The police brought prisoners in on Saturday to hose and scrub the market. "My job was to clean out the stall," recollected Battistella. "It was a dirty job.

"Dad would pay us 50 cents or a dollar for both days. We would have to give Mom half of the money, which she put in the bank. After graduation from high school, we got it back. It taught us the value of a dollar.

"The water and fish juice from the hosing ran down Ursulines Avenue in the street car tracks," he went on. "Old man Ted Liuzza, a sports writer for the *States-Item* newspaper, used to raise hell with the authorities about it."

After graduation from St. Aloysius High School in 1942, Battistella joined the navy during World War II. He tried to get in the submarine service, but he ended up being a landing craft pilot in the Pacific Theatre. Battistella chuckled that he never learned to swim and that the Mae West life jackets they were issued didn't work.

After discharge from the navy in 1945, Battistella worked for his father a short time, then accepted a job as a bellboy at the Roosevelt Hotel. But the shoes hurt his feet, so he went back to the fish business and remained there for fifty-five years.

Battistella and his brother Stanley became minority partners with their father in Battistella Seafood. When the elder Battistella died in the mid-1950s, Battistella bought out some of the other heirs with money he saved during the war.

Brother Stanley continued as Battistella's partner until deteriorating health forced his retirement in the late 1970s. Battistella then bought him out and became sole proprietor of the business.

From his home in Metairie, Battistella talked about the way seafood was sold in the French Market from the 1930s to the 1970s. "Probably 70 percent of the businessmen were Italian. There were a couple of Greeks and a few Yugoslavians."

Battistella explained that the French Market had three kinds of seafood dealers. Some were retailers who sold to the public and were typically open from 7:00 A.M. until 7:00 or 8:00 P.M. The market typically had five or six retail stalls. Some retailed many kinds of seafood, but others were specialists. Battistella recalled, for example, that Angeletti Seafood was a shrimp specialist.

Other dealers sold seafood retail and were also seafood suppliers to hotels and restaurants. They opened at 7:00 A.M. for retail sales, but often stayed open very late at night or occasionally all night to process seafood for hotels and restaurants. Battistella remembered that for a period of time his family's firm and Bagille Seafood, first owned by Joe Bagille, Johnny LaNasa, and Alphonse Wegman and later by Bob Wegmann and John Holtgreve, were the only two such dealers located in the French Market.

Last was a form of wholesaler known as a commission merchant. Fishermen would deliver their catch by truck to the commission merchant and then often rent rooms on Decatur Street and get a meal while the commission merchant sold their catch. The commission merchants would open at 4:00 or 5:00 P.M. and often work all night, arguing over the price of the seafood they were selling to retailers and retailer/processors. Then they would deliver the seafood to the purchasing dealers early the next morning. For their efforts, they retained a percentage of the proceeds of the sales. Approximately ten to twelve seafood commission merchants worked in the French Market over the years of Battistella's memory.

Battistella remembered that when he was a youngster, commercial fishermen delivered speckled trout, redfish, and small drum "by the hand." A hand was a stripped palmetto frond on which the fish were strung. Knots were tied on top, one knot for big fish, two for medium fish, and three for small fish.

After the commission merchants delivered the fish to Battistella's, they would sell them by telephone to the hotels and restaurants or their buyers would come to the market and pick out what they wanted. The fish were delivered to the buyer the next day.

Throughout the process, the fish and seafood were kept iced with ice from an ice company on Chartres Street. The ice was delivered in blocks, then chopped with an axe, and finally crushed with the flat side of the axe.

In the early years, when Battistella was a youngster, deliveries were made with a horse and wagon, but hard-tired trucks rapidly replaced the horses; although Battistella remembered that during the Korean War, when gas and tires became scarce, seafood was again delivered by horse and wagon.

Battistella noted with bemusement that the fish delivery truck in the late 1930s and early 1940s was also used by his father to take the family on Sunday drives. Benches were put in back for the children, and they dropped the curtains if it began raining. "We would go as far out as Gentilly. To go out as far as where the Michoud facility is now was a big trip. It was country then."

In the French Market years, as later on, finfish was the biggest part of Battistella's business. Their mainstay species were speckled trout, redfish, pompano, red snapper, flounder, and sheepshead, with some mullet, channel mullet, and croakers also being in the mix.

They handled no freshwater fish. The pompano and red snapper were not locally caught. They, along with spiny lobster, came in by train from Florida. Battistella's cleaned all finfish to order, whether for retail sale or for delivery to hotels and restaurants.

Except for a short period, no oysters were shucked at Battistella's, but the firm did sell shucked oysters from oyster houses by the gallon and unshucked oysters by the sack.

Live crabs were delivered by truck to the market and sold in wooden bushel baskets. They preferred to sell shrimp head-on but often had to dehead them after two or three days to keep them from going bad.

A few crawfish were sold unenthusiastically. Battistella dumped the crawfish, which were delivered in onion sacks, into wire cages and sold them by the pound. "If you held them too long, you got too many stiffs."

Buyers preferred to buy shrimp in head-on form in the French Market. (Courtesy William D. Chauvin Collection, Louisiana State Archives)

No cooking was done in the Battistella stall, but they did sell some oddities. "We sold a lot of rabbits," said Battistella. "Saturday was a big day. Sometimes we would be knee deep in rabbit parts by the end of the day."

The rabbits were shipped in, simply gutted, by train from Texas. "The shippers would put ice in a burlap sack and stand the sack up in a wooden flour barrel. The rabbits were packed with their bellies toward the ice in the sack."

Battistella took pains to explain that flour barrels had wooden bands and gaps between the slats. This allowed for drainage. Metal-banded wooden wine barrels, which were "tight" and wouldn't leak, were used for seafood, not rabbits.

With obvious relish, Battistella talked about another specialty item they handled—turtles. "Turtles were sold as 'cowan' in New Orleans. They were seasonal but a big item. We butchered loggerheads (alligator snapping turtles) and sea turtles and sold the meat by the pound. A two-ton truck would come in loaded with sea turtles. They would just drop them in the street, and we would crack their heads with axes. Then we iced them overnight.

"Sea turtle meat was good. The meat near the shoulder was a

Sea turtles were a popular seafood item for Battistella's Seafood in the French Market. (Courtesy Byron Despaux)

delicacy. Mom would bread it and fry it. Whew, it was good," Battistella's eyes sparkled.

He went with a topic he enjoyed. "We would get covered with turtle blood and walk into Morning Call [Coffee Stand] just to be ornery. The coffee shop owner would chase me and Gus the Greek (Gus George) out of the shop."

One kind of business that Battistella didn't like was selling small shrimp to sports fishermen when they were open all night. "I hated that! You had to argue over price—two pounds for 25 cents—with people who could afford nice boats and vehicles."

By the 1960s, Battistella estimates that 35 percent of his business was shipping seafood out of the state in wooden barrels and boxes by rail at Union Station. When the trains stopped carrying seafood, a move Battistella said really hurt the seafood business, they shifted to shipping by air. To meet the requirement for leak-proof containers, Battistella pioneered the Batt-pack, a reasonably priced container made of heavy cardboard.

But the French Quarter was becoming congested and getting trucks in and out for shipments became a problem. By the early 1970s, Battistella was gone from the French Market to the Faubourg Marigny.

Battistella (center) with his management team Frank Zuccarelli (left) and Harlon Pearce (right) with their display promoting shark consumption at the Louisiana Restaurant Show, circa 1977.

The modern history of the company had its share of ups and downs. On the positive side, Battistella's Seafood became accepted, grudgingly by its competitors, but still accepted, as the dominant fish house in south Louisiana.

Battistella is proud of how his company developed the food market for sharks, an underutilized species in the mid-1970s. They taught fishermen how to properly handle them, and they created the market for shark meat.

Paul Prudhomme's culinary revolution created opportunities that Battistella took advantage of. "People became willing to try new things," said Battistella "and that created opportunities. Restaurants began to handle nontraditional species. We sold a lot of tuna and other products.

When we first tried to get people to try salmon, we had to really push to sell any. By 2000, when we closed, salmon was big. The same with halibut."

One "bright idea," as he describes it, that didn't work out was shipping crawfish to France. He tells the story. "After some initial shipments, we made a huge shipment of thousands of pounds of live crawfish out by air on a Friday."

Domestic French crawfish interests got wind of it and suddenly an inspector in France decided that the shipment needed a health certificate. "I called everyone I knew," said Battistella, "but I lost that deal. They all died."

"I was successful because I was aggressive and tried things others wouldn't. Being in the seafood business is a hard living. You handle a lot of bucks, but few stay with you."

Probably the biggest blow to the business, said Battistella, was the declaration of redfish as gamefish, followed by the gill net ban. "We had to struggle to find fresh product to sell. We had to look to imports," sighed Battistella.

"In a way though," Battistella mused, "the business hasn't changed much. It is still a fresh market. Some seafood dealers now sell spaghetti, chocolates, and desserts—only a few still cut fresh fish. A lot handle frozen stuff."

He goes on, "The day of the fisherman, the way we did it is long gone. The day of the wild fisherman is gone. We have to do more with aquaculture. Not at sea, but land-based aquaculture.

Prior to their designation as game fish, redfish were one of the key species Battistella Seafood processed for restaurants.

"Few young people are coming up to fish. They go to college to not work so hard. Aquaculture has to be the way. There are plenty of fish but no one to go catch them. There are so many laws and regulations. And fuel costs are high.

"Seafood buyers are different today," adds Battistella. "Major hotels are coming in with centralized purchasing. We used to deal with chefs that you could depend on. Now you have to deal with purchasing agents who are only interested in the bottom line.

"To them it's more important than quality. They will move from dealer to dealer based on price alone. On the supply end, food service distributors are crowding out independent businesses. My biggest thrill," he said, "was to successfully develop accounts by personally working with chefs on an individual basis."

On the subject of the future of post-Katrina New Orleans, a troubled look passes over his face. "New Orleans is recovering quickly. Tourism is increasing. We have a good sports business with two pro teams.

"Tourism and conventions will be back as long as we get good press. Without tourism, you can just put the lights out. Tourism is our backbone. We don't have the labor force needed for big industry.

"How will fish fit in?" He leaves his own question unanswered.

With a resigned voice, Battistella said, "There is more racism in New Orleans now than before the storm. The white community is fed up with the black community—due to racism. And leadership is not doing anything to help the black community.

"The people are what made New Orleans great before, I'm glad I'm on my way out and not on my way in. I had a good time coming up. I have no regrets."

America's Oldest Market

New Orleans is an old city, founded by the French in 1718 on a great bend in the Mississippi River. The river, the reason for the city's founding, was also its supply line. During the period of French governance, food and other items were bought and sold on the river's banks. In 1769, the Spanish took permanent control of the colony, which had been ceded to them six years earlier.

Between 1779 and 1782, the colony's Spanish officials, in an attempt to control prices and improve sanitation, built the first covered market building at the foot of Dumaine Street. Six years later, the city was decimated by a massive fire, and the market was rebuilt in 1791 at the site marked "Old No. 3" on the French Quarter street plat. This building was replaced in 1809 and promptly destroyed by a great

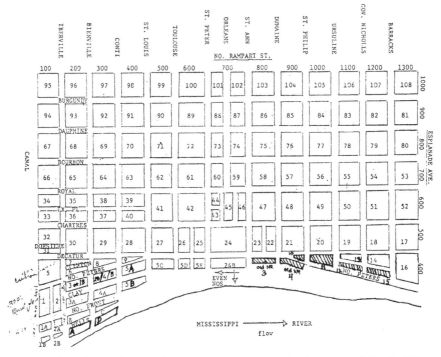

A street plat of the historic French Quarter shows the location of the buildings (shaded) that once made up the French Market. (Courtesy the Historic New Orleans Collection)

hurricane in 1812. It was replaced again the next year by a building, which became known as the Meat Market, Beef Market, or Halle des Boucheries. Before the construction of the Vegetable Market, vegetables and fish were sold here, as well as meat.

The next building constructed, labeled "11" on the plat, known variously as the Vegetable Market, Halle des Légumes, or Marché aux Legumes, was built in 1822 and 1823. This became home to most of the fish and seafood dealers in the market.

"Old No. 4" held the Bazaar Market and the Red Stores. The Bazaar Market, built in 1870, did not house food venders, but rather ninety-three stalls for sellers of dry goods. The Bazaar Market was badly damaged by a hurricane in 1915 and continued in bad condition until 1937, when it was razed to make way for a new arcaded Vegetable and Fruit Market.

At the eastern end of the block were Red Stores Numbers 1, 2, and 3, which are probably best described as "general merchandise stores." Originally built in 1833 by private owners, they were not part of the publically owned French Market. Fire destroyed one of the three stores in 1940, which was rebuilt in a different style. Not surprisingly, the stores were red in color.

The Fruit Market was located on Block 10. In 1847, a waterworks building was built on the small triangular spot off the western end of the Vegetable Market, separated from it by St. Philip Street. This building was adapted as a fruit market in 1853. An article in the August

The Halle des Boucheries was built to house the city's meat suppliers. George François Mugnier. (Courtesy Collections of the Louisiana State Museum)

The Vegetable Market held vegetables and fruits as well as seafood, 1905-10. Alexander Allison. (Courtesy Louisiana Division/City Archives, New Orleans Public Library)

The Bazaar Market was the only building to house venders of non-food items. The Red Stores can be seen on the far end of the market. George François Mugnier. (Courtesy Louisiana Division/City Archives, New Orleans Public Library)

9, 1859, *Daily Picayune* newspaper noted that prior to this "most fruit sellers had only shelves, made fast to the simple iron railing that then surrounded Jackson Square."

The final component of what was considered part of the total French Market, the Farmers' Market (Blocks 12 and 15), did not come until later. A study by the Public Belt Commission in 1924 recommended such a facility, but it was not constructed until 1937-38, during the major rehabilitation of the entire French Market.

The framework set by the French Market as a public market spread rapidly in New Orleans, until virtually every neighborhood held a miniature version of it. The last of the thirty-four neighborhood markets in the city was constructed by 1911, but by that time, some of the older ones had been closed

By city ordinances, these markets held a virtual monopoly on food distribution until the Civil War. After the war, the city allowed the opening of private markets, stores, and stands. Then in 1900, the city reversed itself and made it illegal for private markets within nine blocks of any market to sell fresh fish, meat, or vegetables (except potatoes and onions).

But nothing stemmed the slide of the public markets, including the French Market, into decrepitude. Most had no hot water or toilet facilities and garbage was scattered throughout the markets. The

The Farmers Market, which allowed farmers to sell their produce direct to the public, was the last component put in place in what is traditionally viewed as the French Market.

Private grocery stores offered food at more convenient and sanitary locations and were often less expensive than public markets. This ad was in the July 12, 1935, Times-Picayune *newspaper.*

public, with better transportation available, was willing to travel farther to private markets.

Still, the city was unwilling to abandon the public market concept and began to rehabilitate them. The effort in the French Market, using federal Works Progress Administration (WPA) funds, took place from 1936 to 1937. The Red Stores and the Bazaar Market were demolished, and in their place were built a new vegetable market and a new wholesale fish market. Wholesale seafood dealers were known as "commission merchants," middlemen through whom seafood catches were marketed to other buyers.

A farmers' market to allow farmers to sell their produce directly to the public was built between Ursuline and Barracks streets. Retail fish marketers were moved to the Meat Market. This building and the Vegetable Market received face-lifts, and the entire area was sanitized. The effort rejuvenated the French Market but did little to stem the collapse of the other markets, all of which were sold at public auction within ten years.

After the "Glorification of the French Market," as the project was called, the rejuvenated French Market then began another slow downhill slide. An undertaking to renovate the market by the French Market Corporation through the French Market Complex, an entity managed by Clay Shaw (of Kennedy assassination conspiracy fame) began in the early 1970s.

The food heritage of the market was de-emphasized in favor of enclosed shops, which sold clothing and gifts; restaurants; and, in the former Farmers' Market, a flea market selling imported trinkets. The wholesale seafood market was converted to the "Cuisine Market" to hold restaurants. The produce market on the site of the old Bazaar Market was converted to boutiques. The old Vegetable Market, or Halle des Légumes, was converted to restaurant and retail space. A complex of new Red Stores was built across from the old Vegetable Market.

Traditional businesses fled en masse, nearly dealing the French Market a death blow. By December 1972, only two seafood dealers were left in the market. James T. Georgusis operated his retail fish stall, "Jimmy George's Seafoods" in the Halle des Boucheries. Louis M. Cognevich, the other seafood dealer, was the last fish wholesaler in the market. Cognevich spoke to a *Times-Picayune* newspaper reporter that year.

> We wholesalers heard more about the planned renovation through the newspaper than we did from the French Market Corporation. They told us it was only a question of time before we would be moved. They told

Jimmy George's Seafoods was the last retail stall in the French Market. Its owner, James T. Georgusis, and his father, Hercules, and brother, Gus, were all seafood commission merchants in the French Market in earlier years. (Courtesy William D. Chauvin Collection, Louisiana State Archives.

Battistella's Restaurant in the Vegetable Market, here seen in 1934, was a fixture for more than a half-century. (Courtesy Vieux Carré Survey at the Historic New Orleans Collection)

us they would find a place for us but then we didn't hear anything else. Nothing. Some people got eviction notices, some didn't, so everybody just moved to different places throughout the city. It's only a matter of time before I will have to go too. I'm getting up in age, I'm not too well. Maybe I just ought to retire.

The change affected other businesses as well. The French Market held two famous coffee stands that served thick and chicoried coffee and beignets, Café Du Monde (established in 1862 in the Halle des Boucheries) and Morning Call (established in 1870 in the Halle des Légumes). In 1972, Alvin P. Jurisich, one of Morning Call's owners, was hopefully optimistic that the renovations would result in more business. By 1974, the business had moved to Metairie in the suburbs.

Other casualties occurred. Probably best known was Battistella's Restaurant. The restaurant, owned first by Andrew Battistella, and later his daughter Viola Cristadoro, was best known for adamantly contending that it invented New Orleans' signature sandwich, the poor-boy.

In a story repeated multiple times in the press of the day, including one in the October 6, 1974 in the *Times-Picayune* newspaper, she said:

> I have heard my father tell the story many times about the shabby old Negro man who came into the restaurant one day and begged for something to eat. My father told one of the waiters to "give that poor boy a sandwich," and they made him a huge one—thick slices of ham and cheese with lettuce and tomatoes, on French bread.

"And it was always called a 'poor-boy—never a 'po-boy,'" she added. "He held a copyright on the sandwich for many years, and it was only after he let it expire that others in the city began advertising their own versions."

Cristadoro said that her father began the restaurant in the early 1920s for longshoremen from the river and farmers coming in to the market from the country. It catered to these, but soon picked up a clientele that included "struggling artists, shopkeepers who worked nearby and many Vieux Carré residents."

The restaurant was closed by the French Market Corporation to renovate the building in 1973. The decision not to reopen the restaurant was made in 1974, when Cristadoro found that her rent would be nearly doubled, her space cut in half, major changes in the interior would have to be done at her expense, and the French Market Corporation expected her to spend 4 percent of her gross income for advertising.

"Longshoreman, French Quarter residents and hippies usually don't go to places where prices are for tourists," she said in resignation. "Maybe it will be beautiful when it's all finished, but it won't be the same French Market anymore—at least not for me nor for many other people in the city who knew it in the old days."

Finfish Product Forms

Finfish can be purchased in a variety of product forms or cuts. Most fish species are available in all cuts, but there are exceptions. For example, it would be very difficult to cut a flounder into steak form because of its severely flattened cross-section shape. Below are the five most common cuts with a note on the minimum amount of each to purchase per recipe.

Whole: These are the fish as they come from the water. Before cooking, the fish must be scaled and gutted. Buy ¾ pound per serving.

Drawn: Whole fish with the entrails removed. These must be scaled and de-headed before cooking. Buy ½ to ¾ pound per serving.

Dressed: These fish have the scales, head, and entrails removed. The fins and tail may or may not be removed. These fish can be cooked whole, filleted, or steaked. Buy ½ pound per serving.

Fillets: These are the sides of the fish cut lengthwise away from the backbone. They are all meat and are ready to cook as purchased. Buy ⅓ pound per serving.

Steaks: These are cross-section slices cut from large dressed fish. They are generally ⅝ to 1 inch thick, and the only bone they contain is a cross-section of the backbone and perhaps ribs. Buy ⅓ pound per serving.

Getting Skinned on Fish Names

What's in a name? About seven bucks a pound—when someone sells you Pacific rockfish (a cheap fish) labeled as red snapper (an expensive fish). A few unscrupulous suppliers, retailers, and restaurants do this out of pure avarice. They buy the fish under the less expensive name; price and sell it as more popular, pricey species; and make a ton of money.

Others invent whimsical names (that can resemble other fish) for a product so that they can more easily sell it to people who are cautious about trying new species. An example is selling sheepshead (a delicious table-fish, but with an off-putting name) as "bay snapper." There is no such thing as "bay snapper," but it sounds good.

Either way, the U.S. Food & Drug Administration (FDA) considers such practices to be economic fraud. Sellers are deliberately attempting to deceive consumers while separating them from their money. The FDA is the agency responsible for determining and policing the nomenclature of fish and shellfish.

Sheepshead are a member of the porgy or seabream family and may be sold as "rondeau seabream," but they cannot legally be sold as "bay snapper," a fictitious and misleading name.

Some name problems arise when the accepted common name is just too icky to be able to sell. Would you buy a ratfish, a bloater, a slime flounder, or a bastard halibut? Not likely. Another problem is when a new fishery develops for a species that really has no market name.

The FDA actually is not inflexible, as long as name changes are not attempts to defraud and if the changes receive their blessing. In the late 1970s, a new fishery developed for an unusual deepwater fish off Australia and New Zealand. Millions upon millions of pounds of these delicious fish have been imported and sold in the U.S since then. It was named the "orange roughy." Before that, it was generally known as a "slimehead." In spite of how good it tastes, could a market have been developed for slimehead? Again, not likely.

Closer to home, in 2004, the FDA approved a market name change for the sheepshead at the request of the Louisiana State Seafood Industry Advisory Board. After several months of written negotiations and justifications, the FDA accepted "rondeau seabream" as a legal market name for the species.

The FDA publishes "The Seafood List: FDA's Guide to Acceptable Market Names for Seafood Sold in Interstate Commerce" on their website. This guide lists the approved market names, common names, scientific names, and vernacular names for several hundred species of finfish and shellfish sold in the United States.

What Is Surimi?

Technically, surimi is the raw material that imitation crabmeat, scallops, shrimp, and lobsters are made from. In everyday usage, most people use the term surimi to apply to the imitation shellfish product itself.

Surimi is now a major U.S. product, but it was originally a Japanese product made from highly processed, minced finfish. Most surimi made in the United States uses Alaskan pollock as a raw material. Many surimis have some shellfish in them, so it is important for people with seafood allergies to read the label.

Alaskan pollock used in surimi is partly processed at sea, right after being caught. The bones, skin, and fat are removed, and the flesh is reduced to a white, rubbery paste that is frozen in twenty-two-pound blocks. When the blocks reach the surimi plant, they are thawed and thrown into a giant blender with a variety of additives—salt, sugar, egg whites, starch, polyphosphates, and perhaps crab juice or some

No matter how much they taste or look like crab, surimi products are made of highly processed finfish.

crabmeat or shrimp. Monosodium glutamate and vegetable oils may also be added.

This mixture is shaped into thin sheets or ribbons. These are cooked and cooled, then machine-cut into thin strands. The strands are braided into thicker ropes and dyed red on one or more sides. After dying, they are cut into sticks or chunks for packaging.

Surimi products are fine in their own right but should not be compared to real crab, shrimp, or lobster. A comparison of nutrition labels will show that surimi does not match the nutrition of the products that it imitates.

Most of the surimi products sold in the U.S. are kamaboko-derived, which is surimi cooked by steam. Other surimi types are chikuwa, which is broiled, and satsuma-age, which is fried.

Surimi products are seldom used in top-of-the-line Louisiana restaurants, but they turn up with increasing frequency in home-cooking recipes. Louisiana cooks are surprisingly creative and sometimes produce very good dishes using surimi products.

Surimi products do not freeze well, as they tend to lose a lot of their moisture when thawed.

Louisiana's Caviar

No other state in the U.S. offers both the abundance and variety of edible finfish that Louisiana does. From the cornucopia tumble more than a hundred freshwater and saltwater species, including a unique-to-the-state caviar product, choupique caviar.

Properly defined, caviar is the salted eggs of sturgeons, most often beluga, osetra and sevruga from Iran and Russia, and increasingly kaluga from China. Some farmed French baerii and American white sturgeon caviars are also available. World and U.S. demand for caviar exceeds supply so caviar producers have turned to other fish species with varying success.

Many of these "caviars" do not have the gray to blackish color of true sturgeon caviars, and while their quality may be acceptable, they are not considered good enough to substitute for sturgeon caviar. Among the golden to orange and red roes are salmon, trout, whitefish, flying fish, and lumpfish. Black lumpfish caviar may be purchased, but it has been dyed. The true color of lumpfish roe is orange-red.

Louisiana bowfin caviar is a highly sought after product.

The caviar products closest in quality to sturgeon caviar come from two other fishes as primitive as sturgeons—paddlefish and bowfin. Best is bowfin, and Louisiana is the sole producer of bowfin caviar.

Most commonly known as chopique (pronounced shoe-pick), but also called grinnel, dogfish, and cypress trout, this freshwater species spawns in mid-winter, so fresh bowfin caviar is most available from November through February.

Receiving high reviews by most caviar epicures, sometimes even higher than beluga receives, bowfin caviar is dark gray in color, nearly black. Its turgid berries produce the same delightful, crisp pop with light tooth pressure that good sturgeon caviars produce. And like sturgeon caviar, its taste resembles a faintly briny ocean breeze.

All caviars, including bowfin caviar, should be eaten as simply as possible, without grated onion, capers, or chopped hard-boiled eggs. They may be served on very lightly toasted bread pieces, lightly buttered plain crackers, or small buckwheat pancakes called blinis. Bowfin caviar is typically packed in four-ounce containers, and each ounce will produce eight to ten half-teaspoon servings.

Bowfin caviar is best stored at 28 to 32 degrees Fahrenheit, temperatures difficult to obtain in a home refrigerator. A good method of storage at home is to nest the tin in crushed ice in a bowl, which is

Bowfin, almost always called "choupique" in Louisiana, is a primitive fish like sturgeon.

then placed in the coldest part of the refrigerator. Meltwater should be drained off and the ice replenished every other day or so. Good-quality caviar may be kept this way for several weeks.

Caviar of any type is best not frozen, as freezing makes it softer, affecting its mouth-feel. If it must be frozen, it should be used as soon as possible. Long storage can cause it to develop an objectionable fishy taste.

Caviar should be very gently spooned from the tin, as the roe is delicate and has a fragile skin. Legend has it that caviar cannot be served with a metal spoon of any kind because contact with any metal will impart an objectionable taste. This is completely silly since caviar is always packed in metal tins and held in them for weeks.

Louisiana bowfin caviar can be challenging to find, as never more than six packers produce the state's limited production and the season is short. It may most easily be purchased from wholesale seafood houses that supply New Orleans-area restaurants. Most of these firms will sell certain items at the retail level, as well as wholesale.

Country of Origin Labeling

An ever-larger share of the finfish eaten in the U.S. is imported. Much of this is due to restrictive harvesting laws in this country that have been deemed necessary for conservation. Harvest reductions of varying severity have been put in place for all sharks, most snappers including red snapper, all groupers, cobia, many jacks, king mackerel, bluefin tuna, and swordfish in the Gulf of Mexico.

Also, the commercial fishermen's (and seafood consumer's) share of the wild fish harvest, especially in the southeast, has been significantly reduced. This has been done by legislative action and ballot initiatives that reallocated much of the resource to recreational fishing interests.

Louisiana is a typical example. In 1987, all species of sailfish and marlin were declared off-limits for commercial food harvest by the Louisiana legislature. In 1988, the legislature halted commercial fishing for redfish (red drum).

Then, under intense lobbying by recreational fishing interests, the Louisiana State Legislature in its 1995 session passed comprehensive legislation halting all use of gill nets by 1997, except for strike nets to target pompano and striped mullet (which is seldom consumed in Louisiana). Hook-and-line and trawl harvest was allowed, but the measure almost destroyed consumer supplies of speckled trout and severely impacted black drum, sheepshead, and flounder harvests.

Domestic finfish aquaculture (fish farming) has not been much of a boon for supplies of finfish in the U.S., except for catfish, salmon, and some tilapia. Finfish aquaculture has advanced rapidly in other countries however, particularly in Asia. Many of these farm-raised fish are being shipped to U.S. markets. Large quantities of foreign-caught wild finfish are also being sold to U.S. consumers.

Where fish are caught or raised is important to many consumers. Banned antibiotics and additives have been found in some seafood imports. Some people prefer to "think global and eat local," feeling that long distance transport of food increases its carbon footprint on the earth. Others are patriotic and prefer "made in America." And a lot of people just feel that local products—fruits, vegetables, and seafood—just taste fresher and better.

Whatever one's reasons, the passage of country of origin labeling (COOL) requirements for fresh and frozen seafood by Congress in

Properly labeled fish, such as these, should not only include country of origin, but whether the fish is wild-caught or farm-raised.

2002 has made finding out where seafood comes from easier. Under the law, retailers, except for seafood specialty stores and restaurants, which are both exempt, must inform consumers by a clear label or sign where the product is from and whether it is wild-caught or farm-raised.

To be labeled as a product of the U.S., farm-raised seafood must be hatched, raised, and processed in the U.S. Wild-caught seafood must be harvested from U.S. waters or by a U.S. flagged vessel and processed in the U.S.

Unfortunately, quite a few exemptions from COOL do exist. Cooked, cured, smoked, canned, or surimi-type products are exempt from labeling. When two or more different seafoods are mixed together, they don't need to be labeled, even if they are all imports. Also exempt from labeling requirements are substantially modified seafoods such as breaded products, marinated products, soups, sauces, seafood salads and cocktails, sushi and pâté.

Recognizing Quality Fish:
The Nose Knows

Odor is one of the most useful ways of judging freshness in fish, so indeed "the nose knows." Fresh fish should never smell "fishy." A fishy smell is the first step on the road to putrefaction. Fish should have a neutral, seaweed, or ocean smell. Odor can be used to judge the quality of whole fish as well as any of the unfrozen products from fish, such as fillets and steaks. Never be bashful about asking the salesperson to allow you to smell the fish. They are used to the question coming from educated consumers.

Whole fish can also be assessed for quality by signs other than odor:
- The eyes should be bright, clear, full, and often protruding. Reject if the eyes are dull, cloudy, opaque, and sunken.
- The skin should be bright, with a metallic luster, and markings should be very clear. Reject if the skin is dull, the color faded, and the natural markings are obscured.

Fresh fish have very clear and protruding eyes.

The older a fish is, the cloudier its eyes become and the more they recede into its head.

- The gills should be bright red to slightly pinkish red and have little to no slime. Reject if the gills are gray, greenish, or brown and have a heavy mucous coating.
- The flesh consistency should be firm and spring back elastically when poked with a finger. Reject if the flesh remains dented after poking or if flesh is soft and flabby.

Quality signs also exist for judging the freshness of unfrozen fish that have been cut into fillet, steaks, or dressed fish, but they are more subtle. Odor is the single best indicator, so ask to smell the product. Other indicators include:

- The flesh color should be somewhat translucent and bright. Reject or at least suspect the product if the flesh is very opaque and dull. Reject it if the color is shading to tan or brown.
- The margins of the flesh should be intact and smooth. Reject it if the flesh appears tattered and small strands hang from the cut edges.
- The flesh should be intact, with no gaping or separation of the individual flakes of flesh. Reject it if splits or holes are present between flakes of flesh and segments of the flesh appear to be separating.

Fresh fish have bright red gills.

As fish age, their gill color goes from red to pink to gray to brown.

The flesh of high-quality fish fillets is translucent and bright and the margins of the cut are smooth. The flesh has no gaping between the segments of the fillet.

Frozen fish products are the most difficult to judge for quality. Odor cannot be used, but some other indicators do exist:

- The flesh color should be bright and translucent. Suspect it if the flesh color is dull and opaque and reject it if it is tan or brownish.
- The consistency of the flesh should appear to be intact and not ragged. Reject it if gaps or slits are visible between the flesh flakes or segments or if the cuts appear ragged.
- The flesh should have no white "frosty-looking" spots on it. Reject if frosty spots indicating freezer burn are present.
- The packaging should have no ice crystals inside. Reject it if moisture loss from the fish is indicated by the presence of ice crystals inside the packaging.

Spaghetti Worms in Fish

Spaghetti worms got their name because they look like a crumpled piece of cooked thin spaghetti. They are common parasites of saltwater fish in the drum family, which includes speckled and white trout, black drum, redfish, and croakers. While they look alike to most fishermen, several different worms use these fish as hosts. Most common in sea trout is *Poecilancistrium caryophyllum*. Worms found in black drum are most often *Pseudogrillotia pleistacantha*. For ease of discussion, we will dispose of these tongue twisting Latin names and refer to them all as spaghetti worms.

Fishermen frequently find these white, one- to three-inch-long worms when filleting their catch. In trout, they are usually found in the middle of the fillet in the area just below the dorsal fin. Research has shown that approximately 40 percent of Louisiana and Mississippi speckled trout are host to spaghetti worms, with an average of between one and two worms occurring per fish. It may appear that many more worms exist, but often one worm is cut into several pieces during filleting. Spaghetti worms in black drum are more common near the tail of the fish with a typical fish hosting five to fifteen specimens.

Large black drum are a common host for spaghetti worms.

The spaghetti worms we see in these fish are really parasitic tapeworms of sharks that are just using the trout or drum as an intermediate host. The cycle begins with eggs produced by an eight-inch-long adult worm that lives in a shark's intestine. After being passed into seawater, the egg hatches into a tiny swimming larva called a coracidium. If this larva is eaten within two days by a small marine crustacean like a copepod, it develops into another stage called a procercoid.

At this stage, some uncertainty exists as to what happens. The copepod may be eaten by a trout, passing the larval worm on to the trout. However, since small animals like copepods are seldom eaten by larger trout and very few trout under ten inches long have spaghetti worms, another host is suspected. More than likely, a small bait fish eats the copepod and then it in turn is eaten by a trout. In any case, once the larval worm is in the trout's digestive tract, it tunnels its way into the trout's flesh, where it may live for several years. The life cycle is completed when a shark eats the trout and then serves as host for the adult worm.

The number of trout carrying worms seems to be directly related to the characteristics and quality of the water in which the trout live. In general, the saltier the water and the less polluted it is, the higher the incidence of infection is. This may be due to either one of the

Spaghetti worms are highly visible in raw fish flesh.

intermediate host's or the larval worm's needs for saline, unpolluted waters.

Another interesting fact is that once a trout becomes host to one or several spaghetti worms, it seems to develop immunity to further infections. If this were not the case, large, old fish would have many more worms than a twelve- or fourteen-inch fish, but they don't.

While the spaghetti worm may be somewhat unappealing to the eye, it certainly doesn't prevent good eating. Since they are large enough to easily see, they are fairly simple to remove during the filleting process. Simply grab the worm between the knife blade and thumb and gently pull it out. With a little practice, it becomes easy.

Many people don't even bother to remove them before cooking. After cooking, they are unnoticeable and cannot be tasted. In fact, they seem to disappear. In a survey conducted at Mississippi fishing rodeos, less than 25 percent of the trout fishermen avoided eating fish with worms.

Cooking does, of course, kill the worm. Even without cooking, though, they are not a human health problem. No human infections have been recorded and researchers have been unable to infect warm-blooded animals with the parasite.

Bon appétit!

Fish Poisoning

Few seafood toxins of concern exist in the northern Gulf of Mexico. The one of most concern, scombrotoxin or histamine, is not naturally occurring but produced by mishandling tuna, mackerel, bluefish, dolphin (mahi-mahi), and jacks, such as amberjack. Bacteria form the toxin from an amino acid when these fish begin to spoil because they were not iced quickly enough after capture. In severe cases, the exposed flesh of the fish may develop a honeycomb appearance.

The histamine-forming process is rapid, and once it has been formed, it cannot be removed or destroyed by cooking. Reactions to consuming histamine resemble an allergic reaction and are seldom fatal. Symptoms, which last four to twelve hours, include skin rashes and swelling, headache, pounding heart, flushing, itching, and intestinal problems.

Histamine is seldom encountered in commercially caught fishes. Recreational anglers should plan ahead to have enough ice and space to keep these fish cold.

Another toxin is ciguatera. Ordinarily a toxin associated with tropical reef fishes, ciguatera was considered to be so rare in fish in the northern Gulf as to be nonexistent. At least, until March of 2007. At this time, a Galveston, Texas, couple were thought to have become ill from eating a gag grouper caught on the Flower Gardens reef, off the Louisiana/Texas border. This was followed by more cases. Sixty or so cases have been linked to fish caught from the Flower Gardens and sold in other areas. Most were in St. Louis, but some were also in Washington, D.C.

Symptoms of ciguatera poisoning include nausea, vomiting, and diarrhea; numbness and tingling of the mouth, hands, or feet; joint pain; muscle pain; headache; reversal of hot and cold sensation (such as cold objects feel hot and vice versa); sensitivity to temperature changes; vertigo; and muscular weakness. There also can be cardiovascular problems, including irregular heartbeat and reduced blood pressure.

Symptoms usually appear within hours after eating a toxic fish and disappear within a few weeks. However, in some cases, neurological symptoms can last for months to years. There is no antidote for ciguatera fish poisoning; symptoms can be treated most effectively

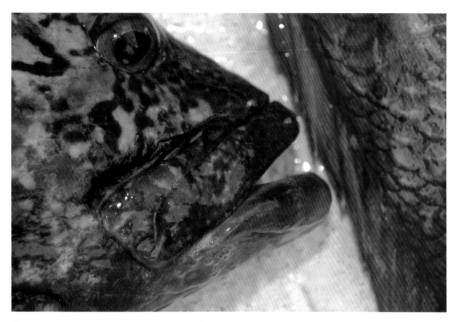

Reef fishes such as these, as well as jacks, were implicated in the 2007 incidences of ciguatera poisoning, the first ever on record in Louisiana.

if diagnosed by a doctor within seventy-two hours. Ciguatera fish poisoning is rarely fatal.

The toxin is produced by a naturally occurring microscopic dinoflagellate, *Giambierdiscus toxicus.* It grows on hard surfaces, so in the tropics it is more common near reefs. As the toxin moves up the food chain in fishes it accumulates to higher and higher levels, peaking in the flesh of large predator fish. Barracuda have been the fish species most often implicated in ciguatera. Other species include various groupers, snappers, jacks, hogfish, and king mackerel.

The U.S. Food & Drug Administration has issued an advisory to seafood processors that ciguatera poisoning was "reasonably likely" to occur from consuming any of several species of fish caught within fifty miles of the Flower Gardens.

Finfish and Food Additives

Finfish are among the least processed of seafoods and therefore have few additives. Most fish are simply cleaned to remove their heads and viscera and perhaps fins and bones, and then sold fresh or frozen. Unlike with shrimp, crabmeat, and crawfish tails, finfish are not cooked in processing. This means that few additives are added to finfish. There are a few, however.

Phosphates, often sodium tripolyphosphate, are applied to catfish that will be frozen. Processors of both farm-raised and wild fish use the product, which is designed to prevent moisture loss from damaged tissue cells when the product is thawed.

Some finfish products may be soaked in ascorbic acid, the stuff of citrus juices, prior to freezing. Ascorbic acid is an antioxidant that prevents fish flesh from becoming darker during frozen storage. After the fish is frozen, it may be further glazed with an ascorbic acid solution. Ascorbic acid is often used on imported fish but seldom on Gulf fish.

Some Gulf of Mexico finfish processors rinse fillets in ozonated

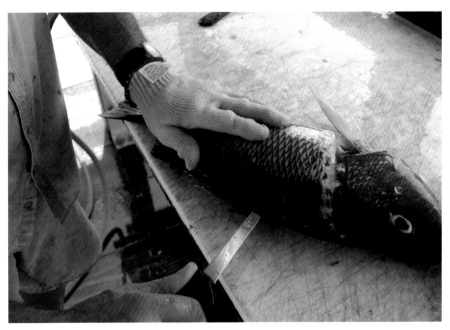

Most finfish processed for sale will contain no food additives.

water. Ozone is simply a form of oxygen molecule in which three oxygen atoms bind to each other rather than two. The normal oxygen molecule is odorless and colorless. Ozone is light blue and has the smell of clothes that have been dried outside on a clothesline.

In spite of its simplicity, ozone is lethal to disease-causing and spoilage organisms such as viruses and bacteria. Fillets soaked in ozone water will, without altering the appearance, color, or aroma of the fish, stay fresh two days longer than untreated fillets.

A controversial treatment method that is approved by the U.S. Food & Drug Administration is a form of "modified atmospheric packaging," sometimes erroneously called "cold smoking." Under the process, fish products are exposed to carbon monoxide (CO) or wood smoke, which has had the taste of smoke filtered out, leaving the CO behind.

Either way, the CO binds with muscle pigments and prevents the flesh from changing color as it ages. For some species, color change is an important indicator of quality. For example, in tuna, as the fish ages, the flesh color changes from true red to brownish-red, to reddish-brown, and then to brown. CO-treated tuna stays hot pink-red in color. White-fleshed fish are also more often being CO-treated than in the past.

Any additives used, including treatment with carbon monoxide (CO), are required to be displayed on the label.

Opponents of CO-treatment say that using CO makes it difficult for wholesalers, retailers, and consumers to judge fish freshness, adding that anything that masks the true age of a piece of fish is a public safety risk. Other detractors of the process say that there is no reason to use CO except to deceive consumers.

Proponents say CO-treatment has been proven not to be a health risk for humans. Other methods, they say, exist besides color to judge the age of fish and an attractively colored product sells faster, spending less time on shelves.

A substantial percentage of unfrozen, domestic saltwater fish are CO-treated, as are a preponderance of imported fish. FDA requires stores to label CO-treated fish. Whether they do or not is another matter.

Storing Fresh Fish

The secret to storing fresh fish until you use them is simple. Use ice, more ice, and then some more ice. This is especially true for whole fish or gutted fish. Ice pulls the temperature of fish down as close as possible to freezing without drying the fish out. Additionally, as the ice slowly melts, meltwater washes the bacteria off the fish and away.

Whether one buys fish or catches their own, fresh fish should be stored the same way. A three-inch layer of ice should be placed on the bottom of an ice chest. This keeps the fish out of any bacteria-laden meltwater. A layer of fish is then laid down one fish deep and without crowding. Another layer of ice is put on top of the fish completely covering them, followed by another layer of fish and then ice. Repeat the process until all the fish are iced. As a rule of thumb, two pounds of ice should be used for every pound of fish.

Properly iced fish that were gutted and stored immediately after catch can last as long as ten to twelve days. Ungutted fish have half the storage life of gutted fish. Unfortunately, when fish are purchased, the buyer does not know how many days have elapsed since the fish were caught, so storage life may be substantially shorter.

Once fish are iced, the ice chest should be kept in a cool spot,

A three-inch layer of ice is put in the bottom of the ice chest, followed by a one-fish-deep layer of fish.

Another layer of ice is spread over the fish.

After the first layer of fish is completely covered with ice, another layer of fish may be added. Fish and ice layers are used alternating until the ice chest is full or all the fish are iced.

especially out of the sun. The drain plug should be opened once a day to allow meltwater, and the bacteria it carries, to drain away.

Storage life of whole or gutted fish can be lengthened by rinsing the fish with a strong jet of clean water before icing to wash bacteria off. Gutting dramatically lengthens shelf life because enzymes from the gut will begin attacking the body cavity wall, and then the flesh, within hours after death.

Some research indicates that the storage life of fish can be prolonged by two to four days by "blanching" the fish before they are iced. Blanching consists of dipping the gutted fish in 190-degree water for two seconds. The process destroys bacteria on the fish, but the heat does not penetrate deeper than the skin layer.

Storage of cleaned fish products such as fillets, steaks, or dressed fish is also best when ice is used. These products are too delicate to put in direct contact with ice and ice water for prolonged periods, however. Cleaned fish should be washed, and then put in polyethylene food storage bags. The bags should be buried in ice in a bowl in the refrigerator. Every day or so, fresh ice should be added to replace what has melted.

Fillets, steaks, and dressed fish that were produced from fish iced

Rinsing fish after gutting them and before icing them will increase their storage life dramatically.

immediately after catch, held no more than three days on ice before cleaning, and then iced in the refrigerator will easily last at least seven days with no appreciable loss of quality.

Ice is the magic bullet. Food science research indicates that fish last twice as long when held at 32 degrees Fahrenheit compared to 37 degrees Fahrenheit. Furthermore, the storage life of fish is substantially longer when they are iced at 32 degrees Fahrenheit than when simply refrigerated without ice at 32 degrees Fahrenheit.

Fresh or Frozen?

One of the biggest misconceptions about buying finfish is that fresh is always better than frozen. Some of that is due to the word "fresh" having two definitions. "Fresh" can be defined as being the opposite of "frozen." It is also defined as the opposite of "spoiled" or "stale." Very often, fish will meet the first definition but not the second.

There is no doubt that the freshest of fish, those purchased very shortly after being caught and properly handled every step of the way, are better than frozen fish. But to find such fish . . .

The very act of freezing does indeed cause some changes. Fortunately, most modern commercial facilities are equipped to minimize those with speedy handling and quick freezing at very low temperatures. Fish handled and frozen properly are of better quality than mediocre or poorly handled fresh (unfrozen) fish. Loss of quality in frozen fish can also occur in properly handled fish simply due to time. Fish flesh is delicate and breaks down easily.

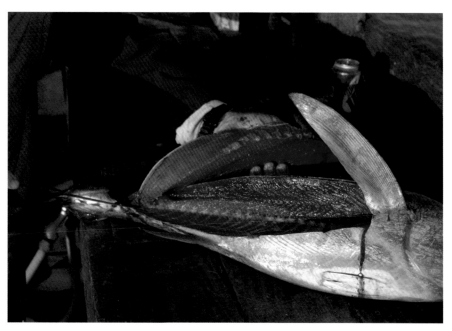

Freshly cleaned fish that have been properly iced and eaten shortly after cleaning are the best on the table.

Improperly packaged fish that have been frozen too long make poor table fare.

Frozen fresh fish can indeed be much better than unfrozen (fresh) fish that isn't fresh.

Finfish and Cholesterol

While confusion existed for years about cholesterol levels in mollusks (oysters, clams, mussels) and the effects of moderately high cholesterol in crustaceans (crabs, shrimp, crawfish, lobster), no such problems existed for finfish. Of all protein sources, they are the low-cholesterol champs.

Only a few finfish species have more than 70 milligrams (mg) of cholesterol per 3½-ounce serving. This compares to 80-90 mg of cholesterol for lean beef, pork, or lamb. Chicken, with or without skin, is only slightly less. Of the major animal protein foods, only mollusks have less cholesterol.

While it makes intuitive sense that very lean, low-fat finfish would be lower in cholesterol than oily, high-fat finfish, that generalization would be incorrect. Dolphin (mahi-mahi) are extremely lean, but have 73 mg of cholesterol per serving. Compare that to pompano, which has 13½ times as much fat per serving, but only 50 mg of cholesterol. Cholesterol levels in finfish are species-specific but are in general quite low and

Whether fat or lean, finfish are among the most heart-friendly of all foods.

fit easily within the American Heart Association's recommendation to limit cholesterol intake to no more than 300 mg daily.

The major exception to the low cholesterol levels of fish is their roe (eggs), including caviar. Finfish roes average about 375 mg of cholesterol per serving, with black and red caviar topping out at 588 mg. Fortunately, these are not everyday foods.

Are Fatty Fish Fattening?

The energy content of foods is measured in calories. Fats are the most calorie-laden food components, at nine calories per gram, compared to four calories per gram for protein and carbohydrates. Obviously, a higher fat fish like pompano at 164 calories per 3½-ounce serving or Spanish mackerel (157 calories) is more "fattening" than flounder with 91 calories or grouper at 92 calories.

But compared to same-sized portions of other protein foods, neither lean nor fatty finfish are really fattening.

Food	Calories per 3½-oz. Serving
Roasted skinless chicken breast	454
Broiled top round beef steak	510
Roasted lamb, lean leg shank	539
Roasted pork, lean leg shank	652
Broiled ground beef, 21 percent fat	930

The decadently delicious pompano is considered to be a high-fat species but is much lower in calories than almost any other animal protein.

Seafood Allergies

Food allergies are caused by the human body reacting to something eaten as if it were a dangerous substance. To combat the perceived threat, the body's immune system releases antibodies into the blood stream. These in turn react with mast cells and basophils in the blood to cause the release of other chemicals, especially histamines in the blood.

These processes cause mild to severe symptoms in the allergic individual, depending on how sensitive the person is. Skin problems, such as itchiness, hives, and swelling may occur, as well as sneezing, runny nose, or sinus problems.

In a more severe reaction, the throat and tongue may swell and tingling may occur in the mouth. Abdominal pain, diarrhea, and heartburn can be other symptoms, as can wheezing and coughing.

A very severe reaction can produce anaphylactic shock, which may be life-threatening. Breathing becomes very difficult and a choking sensation is present. Blood pressure drops and the individual can become unconscious.

A surprising number of foods, like this Worcestershire sauce, contain finfish.

People allergic to one kind of finfish are likely allergic to all species of finfish, as all fish share the protein parvalbumin. Children are more likely than adults to show finfish allergies, so apparently some people outgrow the allergy. However, anyone showing a seafood allergy should consult a doctor before attempting to eat other seafood.

Individuals allergic to finfish may also have other seafood allergies. Crustacean shellfish allergies involve shrimp, crabs, crawfish, and lobsters. People with molluscan shellfish allergies should avoid oysters, clams, mussels, snails, squid, and octopus. Some people have all three seafood allergies, while others may just be allergic to finfish. Raw seafood will usually trigger a stronger reaction than cooked seafood.

The only treatment for seafood allergies is to completely avoid the allergen. This can be difficult for those suffering from finfish allergies. Anchovies are often found as an ingredient in Worcestershire sauce, steak sauces, Caesar salad and Caesar salad dressing, some marinara sauces, and caponata.

Those with finfish allergies, but not crustacean allergies, should avoid consumption of surimi, often called imitation crab or lobster. Surimi is produced mostly from finfish, although some crustacean product is usually present. Surimi, in turn, may be used for flavoring hot dogs, bologna, and ham.

People allergic to finfish rapidly learn to read labels.

How to Fillet a Fish

Fish may be filleted using either an electric knife or a standard steel blade, depending upon the preference of the user. Electric knives are especially useful for hard-boned and large-scaled fish such as redfish (below) but have a few handicaps. They cannot be finessed as well as a steel knife can for cleaning very small fish, such as freshwater bream, or large fish, such as tuna. They also require a source of electrical power and are prone to failure. Having a spare electric knife or a steel blade (and the knowledge to use it) nearby is important. Standard steel blades must be very sharp and therefore require regular sharpening, a skill many have not mastered. The choice depends on the individual. Both tools have strongly opinioned adherents.

Position the fish so that its head may be firmly gripped in one hand and make a diagonal vertical cut down behind the gill covers and down to the backbone.

Still grasping the head, turn the cutting edge of the knife blade towards the tail and cut the length of the fish, keeping the knife in close contact with the backbone. Stop the cut slightly before reaching the tail, leaving a "hinge" of skin and flesh.

Flip the skin-on fillet over and, while holding its base with one finger near the hinge, scrape-cut the flesh of the fillet loose from the skin, beginning near its tail end and ending at the head end.

Remove the rib cage for discard by making a curving vertical cut along its perimeter.

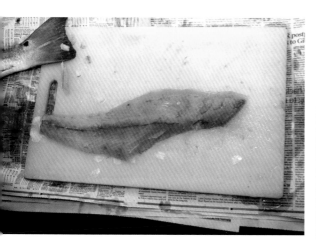

The finished fillet should be checked with fingers along the lateral line (the red line down the center of the fillet in the photo) for any stray in bones. If present, these may be removed with small cut.

Turn the fish over and repeat the process for the other side of the fish.

PART II: RECIPES

Ahi Poke

Poke (pronounced POH-kay) is one of the foods for which Hawaii is known. They even have a poke festival. It's like boudin in south Louisiana—a cult food. Like boudin, every quick-stop store seems to sell it, and each place has its devoted followers. The varieties of poke in the Island State have proliferated, and everything from swordfish through all species of jacks to snapper is used. Tuna is the staple and the best. This is a raw fish preparation, so only the freshest fish—sashimi grade—should be used. If you like sashimi, this will make you melt.

lb. fresh yellowfin tuna (ahi), diced into ¾-inch cubes
tsp. sea salt
2 tsp. pepper
2 cup minced red onion
2 cup minced green onion tops
4 cup soy sauce
tbsp. Asian sesame oil
tbsp. sesame seeds

Tip: Chopped macadamia nuts are a traditional ingredient that may be added, although the dish is wonderful without them.

Place tuna in a bowl and sprinkle with salt and pepper. Toss gently until all of the fish has been seasoned. Add remaining ingredients; toss again to combine. Cover and chill in the refrigerator for at least 1 hour before serving. Serves 4 as a salad, or 6-8 as an appetizer.

Fish Nachos

John Davis's fishing habit provides plenty of fish for the kitchen.

Tip: If serving this dish to kids, you may want to omit the jalapeño pepper.

Tip: The leftover cilantro-ranch dressing may be refrigerated and used as a dip for veggie sticks or chips.

John Davis of Slidell, the originator of this recipe, says that he cooks every day, unless they eat out. "Shanda [his wife] doesn't cook—everybody knows that. She is a big health nut, so I decided to incorporate something healthy into our diet and I personally like nachos. A lot of people put a big clump of sour cream on top. I wanted to change something up, so I made a cilantro-ranch dressing instead. John's choice of fish for the dish is speckled trout, probably because he catches so many at Delacroix, but he has used everything from redfish to store-bought tilapia. Because the fish is broken into pieces, virtually any white-fleshed fish fillet will work, including crappie and freshwater catfish.

Cilantro-Ranch Dressing

8 oz. sour cream
1 cup mayonnaise
1 cup buttermilk
1 package ranch dressing mix
1 bunch cilantro, coarsely chopped

Add all ingredients in a food processor and blend well. Set aside.

Pico de Gallo

4 roma tomatoes, diced
1 small onion, diced
1 jalapeño pepper, diced
¾ bunch cilantro, chopped
1 lime, juiced
Sea salt
Black pepper

In a large bowl add all ingredients and toss to mix. Set aside.

Fish

2 tbsp. olive oil
2 lb. fish fillets
½ cup chopped onion
1 tbsp. minced garlic
1 tbsp. butter
1 1-oz. package taco seasoning mix

Heat olive oil in a large skillet. Add fillets and sauté the fish until it begins to turn opaque. Break fish into small pieces using a spatula. Add onion, garlic, butter, and taco seasoning mix. Stir to blend and continue cooking until fish are done.

Assembling the Nachos

1 package restaurant-style tortilla scoops
8-oz. shredded Colby Jack cheese

Place nacho scoops on a plate, sprinkle with shredded Colby Jack cheese. Spoon fish mixture over cheese. Add fresh pico de gallo over fish. Sprinkle with more Colby Jack cheese and place plate in microwave to melt cheese. Remove from microwave and pour Cilantro-Ranch Dressing over the top. Serves 10-12 as an appetizer, or 4-6 as an entrée.

Labello's Italian Tribute Soup

Jason Labello calls his jambalya pot a baby compared to his father's.

Tip: Jason is very brand specific about a couple of things. First, he uses only Contadina or Del Monte canned tomatoes with basil, garlic, and oregano. Second is his Creole seasoning. He drives one and a half hours each way to Rouse's Supermarket in Thibodaux to get his Benoit's Best Spicy, Salt-Free Cajun Seasoning.

"Did you know that 'Labello' means 'the beautiful' in Italian? That's all the Italian I know though," says Jason Labello of Denham Springs. What he does know is how to cook. His dad, A. J., got Jason into cooking by cooking jambalaya for events. Now, Jason is a member of a Coastal Conservation Association cooking team, does two annual cookouts for his employees at Diesel Specialists, and cooks for family and friends at events. He says that he loves to experiment when he cooks—"a lot." This recipe is "version number 7," according to his count. The idea for the recipe came from an e-mail that he got about a spicy cod chowder. "I didn't have cod, so I used redfish. It was horrible—bland, not interesting— and it had an ugly, average color. I like the idea of soups, and I eat a lot of soups, so I opened my pantry." The result is this—simply delightful. "It's definitely Italian but with a kick," is his description. Jason uses redfish here, but any firm-fleshed fish will work.

1 lb. fish fillets
4 tsp. olive oil
½ white onion, diced
1 tsp. minced garlic
2 14-oz. cans diced tomatoes with basil, garlic, and oregano
1 tsp. dried basil
½ tsp. dried oregano
1 tsp. black pepper
1 14-oz. can chicken broth
3 cups water
1 tbsp. Benoit's Best Spicy, Salt-Free Cajun Seasoning
1 tbsp. onion powder
2 tsp. salt
2 tsp. granulated chicken bouillon
1½ tbsp. Tabasco Green Pepper Sauce
½ tsp. liquid crab boil
2 tbsp. minced parsley

Cut fillets into bite-size pieces and set aside. In a large pan, heat olive oil over medium heat. Add onion and sauté until transparent

ust before the onion is done, stir in garlic. Add tomatoes, basil, oregano, pepper, and chicken broth. Cook uncovered for 2 minutes. Add 3 cups water and bring to a boil. Add Cajun seasoning, onion powder, salt, granulated bouillon, Tabasco, and crab boil. Reduce heat, add fish, and simmer uncovered for 5 minutes or until fish flakes when tested with a fork. Serve in bowls with parsley sprinkled on top. Serves 4-6.

Cajun Cioppino

In 1999, on a visit to the San Francisco Seafood Show, I had the pleasure of eating cioppino, which is considered to be San Francisco's signature dish. Cioppino is to San Franciscans what gumbo is to Louisianans. Cioppino (pronounced chuh-PEE-no) is basically a fish stew. Supposedly, the dish evolved in San Francisco from northern Italian immigrant fishermen. Must-have ingredients include tomatoes, wine, and spices. But the dish is wonderfully versatile in that almost any kind of seafood can be used. A California cioppino will usually use Dungeness crabs, mussels, and clams. We use blue crabs, shrimp, and oysters. The choice of fish to use is endless, just so long as it is a firm-fleshed species that will not cook apart in the liquid. Freshwater fish, inshore fish, and offshore fish work equally well. The last cioppino we cooked used redfish, the one before that used mangrove snapper. You get the idea.

½ cup olive oil
1 large onion, chopped
1 bell pepper, chopped
3 cloves garlic, minced
1 28-oz. can diced tomatoes
2 tbsp. tomato paste
2 cups dry white wine
½ cup chopped fresh parsley, divided
2 tsp. salt
1 tsp. black pepper
1 lb. crabmeat
1 lb. peeled medium shrimp
1½ lb. fish, cut into 1½-inch cubes
3 dozen small shucked oysters

Pour olive oil into a medium-sized saucepan. Add onion, bell pepper, and garlic and sauté until soft. Add tomatoes, tomato paste, wine, half of the parsley, salt, and pepper. Bring to a boil, lower heat, cover, and simmer for 15 minutes. Place the crabmeat, shrimp, fish, and oysters in a large saucepan. After simmering the sauce, pour it over the seafood. If the sauce doesn't cover the seafood, add water

Bring to a boil, cover, and simmer a maximum of 10-12 minutes or until the fish is white and flakey and shrimp are pink. Do not overcook because the fish will fall apart. Serve in large bowls with parsley sprinkled on top. Serves 6.

Miz Glinda's Courtbouillon

Glinda Jenkins cooks huge batches of her courtbouillon and everybody in the neighborhood flocks in to eat.

Tip: The flavor will improve if the dish is made a few hours before serving; just re-heat thoroughly, then add green onion, parsley, and wine.

Tip: Many kinds of fish besides catfish can be used in the courtbouillon, such as red snapper, redfish or gaspergou.

I first ate Glinda Jenkins' courtbouillon after a fishing trip on the Ouachita River in North Louisiana. Duane Taylor and I were famished after having cleaned about 150 pounds of blue and Opelousas catfish from the river. A gallon bag of chunked blue catfish fillets disappeared into Glinda's West Monroe kitchen. Several hours later, this magic dish appeared in front of me. In such a quiet voice that you have to strain to hear, Glinda explains that the base recipe came from a 1975 cookbook called *Mrs. Simms' Fun Cooking Guide,* by Myrtle Landry Simms. Mrs. Simms was her cousin Delores' godmother. Delores sent the cookbook to all the members of the family. Glinda says that she lost her first copy in a house fire and had to get Delores to send her another one. She describes it as a prized possession. Glinda says Mrs. Simms won many awards with her cooking and that Delores followed in her path as an outstanding cook.

1½ lb. catfish fillets, cut in serving-size pieces
Salt to taste
Red pepper to taste
1 clove garlic, minced
¼ cup cooking oil
⅓ cup flour
1 medium onion, minced
1 stalk celery, chopped fine
1 small green bell pepper, chopped
1 small red bell pepper, chopped
4 cups whole tomatoes, chopped
2 tbsp. tomato sauce
1 bay leaf
1 tsp. sugar
3 pints water
Pinch of oregano
Pinch of thyme
½ lemon, sliced
1 tbsp. chopped parsley
1 tbsp. chopped green onion tops
1 jigger sherry

Season the fish generously with salt, pepper, and garlic. Place in a covered dish and chill 8 to 12 hours before cooking. Make a roux by heating the oil in a large heavy-bottomed pot. Add the flour, stir constantly, cooking over a low heat until it is reddish brown. Remove from heat. Add the onion, celery, and green and red bell peppers. Stir constantly for 4 to 5 minutes. Return to heat, add tomatoes, tomato sauce, bay leaf, sugar, and water and stir. When it comes to a hard boil, reduce heat to low, cover the pot with a tight-fitting lid, and simmer for 45 minutes. Add the fish, oregano, thyme, and lemon. Simmer 20 minutes longer. Add parsley, green onions, and sherry, cook 1 minute longer. Serve in soup plates with a scoop of cooked rice. Serves 4.

T-Coon's Seafood Courtbouillon

David Billeaud says this is not your mama's courtbouillon.

Tip: Don't overcook this dish or stir it too often or the fish will break up. As soon as a fork will slide through a bigger piece, turn the heat off. Instead of stirring the dish with a spoon to keep it from sticking, David shakes the pot vigorously a few times after adding the fish.

Tip: Almost any white-fleshed fish can be used in this recipe, but catfish, gaspergou, redfish, and black drum are traditional.

The menu special on Fridays at T-Coon's Restaurant in Lafayette is Catfish Courtbouillon. Owned and operated by the speckled trout fishing fanatic, David Billeaud, the nineteen-year-old restaurant is becoming a local landmark. But the courtbouillon David serves is a lot more than a fish dish. He adds shrimp and crawfish tails, so we took the liberty of changing the name of the dish to "seafood courtbouillon." He says that their addition provides "more surprises for your taste buds." The dish offers other surprises as well. He adds roux to "mellow out" the assertive tomato presence. This courtbouillon isn't red and it isn't brown. Glenda, a tough judge on courtbouillons, calls it just right. David explains that when he was growing up, a traditional courtbouillon was watery, tomato-based, and made in the oven. What he cooks now evolved over years. "Like any other recipe I got, I winged it," he explains. He sits in his quiet restaurant after hours with a bowl of the dish and dissects it as he spoons it into his mouth. "The lemon jumped out at me; the garlic jumped out at me; the onion tops (green onions) jumped out at me. It's a party in my mouth." He grins wickedly.

David has his own version of Creole seasoning called "The Stuff," which he sells at the restaurant. If it is unavailable, use the Creole seasoning of your choice.

1 5-lb. catfish
3 cups water, divided
The Stuff seasoning
2 sticks butter
3 cups chopped celery
3 cups chopped onion
1 tsp. minced garlic
½ cup tomato paste
½ cup roux
1 lemon, juiced
1 tbsp. sugar
1 lb. peeled medium shrimp
1 lb. peeled crawfish
4 green onions, chopped

Fillet the catfish and set fillets aside. Discard viscera and air bladder. Trim fins from remaining frame and chop it into pieces that will fit in a saucepan. Add 2 cups water and 1 tsp. The Stuff. Bring to a boil. Cook for 20 minutes, adding water as needed to make 1 ½ cups fish stock. While stock is cooking, cut the fillets into bite-size pieces. Season liberally with The Stuff and set aside. Melt butter in a large Dutch oven. Add celery, onion, and garlic. Sauté until tender. Strain the fish stock and discard the bones. Add 1½ cups of stock to the sautéed vegetables. Stir in the tomato paste and roux and mix well. Add 1 cup water, 1 tsp. The Stuff, lemon juice, and sugar. Simmer for 5 minutes. Add shrimp, crawfish, and fish. Cover, lower heat, and simmer until seafood is done (about 20 minutes). Do not stir because the fish will break up. Shake the pot periodically to prevent sticking. Sprinkle with green onions and serve. Serves 12-15.

Louisiana Bouillabaisse

Terry St. Cyr of Lafayette prepared this dish for Glenda and me at his Grand Isle camp. He calls it a great camp dish because once it is assembled it almost cooks itself while the bourré game is going on. Terry sent me to Magnus Arceneaux Jr., a family friend from whom he got the recipe. Magnus, who lives in Larose, was born in 1939 and has been cooking almost ever since. Four men in the family have received the name Magnus (and a girl was named Maggie), all descended from Magnus Sr., who was a truck farmer and owned a bus converted to a rolling store in Lafourche Parish. Magnus himself refuses ultimate credit for the recipe, saying that he got it from his father-in-law, Nolan Vinet Sr., a Lafourche Parish oysterman. Magnus has prepared this dish at the Louisiana Pavilion of the New Orleans World's Fair and the French Food Festival in Larose, which was known as the Bouillabaisse Festival until about twenty-five years ago.

2¾ lb. redfish fillets
1¾ lb. peeled small shrimp
Creole seasoning
½ cup olive oil
2 large onions, chopped
½ large bell pepper, chopped
6 tbsp. minced garlic
1 lb. crabmeat
1 10-oz. can Rotel tomatoes
1 cup stuffed olives
2 lemons, sliced

Liberally sprinkle fillets and shrimp with Creole seasoning and set aside. Pour olive oil into a medium Dutch oven and layer half of the onions, bell pepper, and garlic in the olive oil. Next, layer half of the fish, then shrimp, then crabmeat. Repeat layers of each ingredient except olive oil. Pour Rotel tomatoes and olives over the layered ingredients. Spread lemon slices over the top. Turn heat to low, and simmer covered for 1 hour. Serves 6-8.

Ceviche

My first taste of this Latin American fish salad came on a snapper-grouper boat based out of Tarpon Springs, Florida and fishing off the Louisiana coast. Its preparer was a salty old deck hand who was so much at home on the deck of a heaving boat at sea that he had trouble walking on level land. He picked the recipe up while fishing for snappers and groupers out of Mexican and Caribbean ports. I loved it. He used grouper, but any white, flakey-fleshed fish is good, including any of the many species of snappers. But my favorite fish for the dish is freshwater crappie, which I freeze first because of the remote possibility of parasites in some freshwater fish. In ceviche, the flesh of the fish is cooked with the acid of lemons and/or limes rather than with heat. This wonderful dish is excellent for light summertime dining and, with its holiday colors, is a good Christmas appetizer.

2 lb. lean white fish fillets
3 large tomatoes
2 bell peppers
10 green onions
¼ cup olive oil
1 tbsp. dried oregano leaves
1 tbsp. salt
1 tsp. black pepper
10 lemons
10 limes

Cut the fish, tomatoes, and bell peppers into bite-size pieces and place in a glass or stainless steel bowl. Dice green onions and add to bowl. Add olive oil and seasonings. Juice the lemons and limes and add to the mixture. Mix well. If you don't have enough liquid to cover the mixture, add more lemon or lime juice. Set aside in the refrigerator for at least 6 hours, or better yet, 12 hours before serving. May be served with or without crackers. Serves 4-6.

Seared Ginger Tuna Salad

This recipe is ideal for using the small steaks cut from the pointy end of a tuna loin as well as the steaks from the small loins of blackfin tuna. Blackfins seldom grow large enough to produce five- to eight ounce steaks, what are needed for each serving. The marinade is delicious, being perfumed by the heady aroma of fresh ginger. This recipe can be served as a side salad, but it is meaty enough that it is better as an entrée. A medallion is simply a name for a small steak.

Marinade

2 tbsp. dark sesame oil
2 tbsp. soy sauce
1 tbsp. grated fresh ginger
1 clove garlic, minced
1 green onion, thinly sliced
1 tsp. lime juice

Mix all marinade ingredients in a small bowl. Coat the tuna medallions with the marinade. Cover tightly and refrigerate for 1 hour.

Salad Dressing

½ cup olive oil
¼ cup balsamic vinegar
¼ cup fresh lime juice
¼ cup orange juice
2 tbsp. soy sauce
2 tbsp. sesame oil
2 tbsp. minced fresh chives
1 tbsp. minced fresh ginger
¼ tsp. salt
Pepper to taste

In a small bowl, whisk together the dressing ingredients. Set aside

Salad

3 oz. mixed lettuce or baby greens
2 large cherry tomatoes, cut in half
1 11-oz. can mandarin orange slices, drained

Prepare the salad before searing the tuna. Store in the refrigerator until ready to use.

Tuna

1½ lb. tuna medallions (¾ inch thick)
2 tbsp. cooking oil, divided

Add 1 tbsp. cooking oil to an iron skillet and heat over high heat until the oil is simmering and near smoking. Add half of the tuna medallions and cook for 30-45 seconds on each side. Remove from heat and repeat the process with the rest of the tuna. Place the seared tuna on the salad and pour dressing over the salad. Serve immediately. Serves 4.

Pelican Fish Salad

This delightful recipe provides a new option for using fish. We were a little nervous about trying it because cold fish flakes didn't sound like anything savory. But they are. And it is a great meal during the hot summer, when fish bite best, your supply of fresh fillets is largest and cooking hot meals is the least fun. With very little to disguise the taste, the pure, natural taste of the fish is featured, so the fish must be fresh and well-kept. This is not a recipe for frozen fish, as freezing tends to soften the texture of fish flesh.

1 lb. cold fish fillets, boiled and flaked
1 medium green pepper, slivered
1 14½-oz. can cut green beans
1 cucumber, sliced
1 medium onion, cut in rings
1 8-oz. bottle Italian salad dressing
2 medium tomatoes, cut in wedges
Salt and pepper, to taste
Leaf lettuce or spinach (optional)

Mix all ingredients except tomatoes, salt, pepper, and lettuce or spinach in a large bowl. Refrigerate 1 hour, stirring once. Remove from refrigerator and add tomatoes, salt, and pepper immediately before serving. Serve as is or over beds of lettuce or fresh spinach. Serves 6.

Sheepshead is an excellent choice for this dish, although many others may be used as well.

Fried Fish Salad

It's hard to tell whether David Billeaud of Lafayette is better known for his cooking or his fishing prowess. He has won two completely rigged boats in four years of competition in the Coastal Conservation Association STAR fishing tournament. His fishing specialty is for speckled trout. As for cooking, he owns and operates the hugely popular T-Coon's Restaurant in Lafayette. At breakfast and lunch, a line of people can be found snaking out the door of the popular eatery on the corner of West Pinhook and Kaliste Saloom roads. David gives credit for the recipe to his wife Paige's grandmother, Edna Mae Grossie. It's a wonderful way to use leftover fried fish of any type, but it is so good that it is worth frying fish especially to make the salad. Serve it on bread, crackers, or use it to stuff fresh locally grown tomatoes. The addition of shrimp really makes the salad sparkle.

cups coarsely chopped fried catfish
cups coarsely chopped fried shrimp
cups chopped celery
cup chopped onion
boiled eggs, chopped
cup mayonnaise
tbsp. mustard
cup sweet pickle relish

Mix all ingredients well and serve on toast. Serves 6-8 as a meal.

David Billeaud's fishing friend, Harold Schoeffler, with a blue catfish like the one used to make the fried fish salad.

Grilled Tuna Steaks

Tuna steaks are produced by cross-cutting loins removed from the length of the fish.

Tip: Allow enough time to marinate the steaks, but do not over marinate them or the acid will begin to "cook" the flesh.

As hard as it is to believe, as recently as the early 1980s, no one in Louisiana fished for tuna, either recreationally or commercially. And the only tuna eaten was the stuff in the can. Now, with big recreational and commercial fisheries in the state, fresh tuna has almost become a staple item, and grilling it is the gold standard of preparations. Tuna steaks are usually marinated before cooking. The principles of marinades for tuna are simple: an oil, often olive oil; an acid, some sort of vinegar or citrus juice; and finally, whatever seasonings that tickle your fancy. If you want to take a perfectly acceptable short cut to a marinade, simply use an eight-ounce bottle of Italian salad dressing and one tablespoon of liquid smoke.

4 6-oz. 1-inch-thick tuna steaks
½ cup white wine vinegar
½ tbsp. minced garlic
6 tbsp. water
1 cup olive oil
1 tbsp. Mrs. Dash Table Blend
1 tbsp. garlic salt
¼ tsp. dried parsley flakes
¼ tsp. garlic powder
⅛ tsp. onion powder
⅛ tsp. oregano
¼ tsp. basil
¼ tsp. paprika
¼ tsp. celery seeds
1 ½ tsp. sugar
⅛ tsp. pepper
1 tbsp. liquid smoke

Rinse the tuna steaks, pat dry, and set aside in a plastic zipper bag. Mix the vinegar and garlic and set aside. Combine all other ingredients, and then combine them with the garlic and vinegar. Add the marinade to the steaks in the bag and flip over several times to be sure that all parts of the steaks are coated. Marinate in refrigerator for about 1 hour, but no more than 2 hours. Turn bag over to recoat he steaks at least once. Build a hot fire in the barbeque pit. Remove he steaks from the marinade and place them on the hot grating. Grill -3 minutes on one side, then turn over and grill the other side for 2-3 ninutes. Remove from heat and serve immediately. Do not allow the teaks to rest or residual heat may cause them to overcook. Serves 4.

Kayleigh's Stuffed Redfish

If you understand Kayleigh Stansel's credentials, you will understand why this dish is so good. She does indeed hold a degree in culinary arts from Sowella Technical Community College in Lake Charles, but that is the least of it. She grew up cooking and fishing. At age five, her dad, Guy Stansel, taught her how to make an egg sandwich, and she has been cooking ever since.

She started fishing, hunting, alligatoring, and crabbing before then and has done so ever since. And she has a lot of opportunities. Her father and his brothers, Kirk and Bobby, own and operate Hackberry Rod and Gun Club in Hackberry, Louisiana. The guide service and lodge on the shores of Lake Calcasieu has a complete dining room, and guests there often eat Kayleigh's creations. This dish won first place in competition in the Alligator Soiree at the 2009 Louisiana Restaurant Association Show. The recipe is also in the *Hackberry Rod and Gun Club Cookbook*.

Tip: Add enough cracker crumbs to the stuffing so that it can make a loose ball when held in your hand.

Tip: When turning fish with a spatula, place the spatula at the front end of the fillet and slide it in the same direction that the scales lay.

4 18-20-inch skin-on redfish fillets, rib cage in
1 cup Italian salad dressing
1 red bell pepper, diced
1 cup diced onion
1 stick butter, melted
1 lb. peeled shrimp
1 lb. crabmeat
2 tbsp. Worcestershire sauce
Salt and pepper to taste
1 cup cream
3 cups cracker crumbs
1 tbsp. white wine
1 tbsp. Tony Chachere's Original Creole Seasoning
1 tbsp. butter
Juice of 1 lemon

Cut the flesh of each fillet to the skin down the center line for most of its length. Use the knife to cut each side, separating the skin from most of the flesh. Do not cut all the way to the edge of the fillet. The intent is to make flaps that will form a pocket to hold the stuffing.

Marinate the fillets in Italian dressing for 15 minutes. In a large saucepan, sauté bell pepper and onion in 1 stick of butter. Add shrimp and cook for 5 minutes. Fold in crabmeat, Worcestershire sauce, salt, and pepper. Pour in cream and gently stir to blend. Remove from heat. Gradually add cracker crumbs until desired consistency is achieved. Make a loose ball with the mixture and place in the pocket of the fillet. The stuffing should be mounded on top of the fish. In a small saucepan, heat white wine, Creole seasoning, and 1 tbsp. butter until blended. Baste the fish with the wine sauce while grilling. Grill until the fish flesh flakes, then move the fillets away from the coals to finish cooking the stuffing. Drizzle lemon juice over the fillets before serving. Serves 4.

Fish on the Half-Shell with Secret Sauce

Cooking un-skinned fish fillets skin side down on a grill has become very popular in Louisiana. What makes this recipe from David Meyer of Broussard, Louisiana, special is his "secret sauce." When David prepared it for me the first time at his duck-hunting camp, he used red snapper fillets. I've tried it with redfish and it is equally good. David says that any fish with large scales, including any species of snapper, redfish, black drum, tripletail, and even largemouth bass is ideal for this dish. Fish with small scales, such as speckled trout, are a little fragile for the preparation but will also work if care is taken when handling them.

4 ¾-lb. fillets with skin and scales intact
2 sticks butter
1 medium bell pepper, chopped
1 medium onion, chopped
6 cloves garlic, minced
4 oz. Worcestershire sauce
¼ tsp. salt
Creole seasoning to taste

Prepare barbeque pit for grilling. While charcoal is lighting, rinse fish fillets and set aside to drain. Melt butter in a saucepan over medium-low heat. Add bell pepper, onion, garlic, Worcestershire sauce, and salt. Sauté until onions and bell pepper are very soft. Do not caramelize! When the sauce is nearly ready and the charcoal is ready, place fish fillets skin side down on the pit's grates. Season liberally with Creole seasoning. Cook 10-15 minutes or until the flesh is turning white and opaque. Spoon the sauce evenly over the fillets. Grill until tender, approximately 10 minutes. Serve on the hardened skin shell. Serves 4.

Barbecued Fish

This dish is for lovers of red sauces and marinades. The ketchup glazes the fillet beautifully and gives it its distinct tomato taste, without being "ketchupy." Thick fillets, ones too thick to fry properly, are best for this dish. We like to use large sheepshead or bass, but other fillets from large fish will work as well. Soft-fleshed fish like catfish are much easier to cook with this recipe if a folding wire grilling basket is used.

1½ lb. thick fish fillets
½ cup ketchup
¼ cup cooking oil
3 tbsp. lemon juice
2 tbsp. vinegar
2 tbsp. liquid smoke
1 tbsp. minced onion
1 tsp. salt
1 tsp. Worcestershire sauce
½ tsp. dry mustard
¼ tsp. paprika
1 clove garlic, minced
3 dashes hot sauce

Cut fish into serving-size pieces. Combine all other ingredients and stir. Place fish and sauce in a bowl and mix well to spread sauce over all the fish. Marinate for 30 minutes. Remove fish and reserve marinade. Grill over medium hot coals for 10-15 minutes or until fish flesh flakes when tested with a fork or knife. Baste once or twice with marinade while cooking. Serves 4.

Mangroves and Mangos

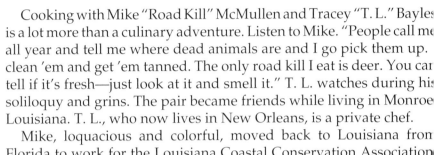

Cooking with Mike "Road Kill" McMullen and Tracey "T. L." Bayles is a lot more than a culinary adventure. Listen to Mike. "People call me all year and tell me where dead animals are and I go pick them up. I clean 'em and get 'em tanned. The only road kill I eat is deer. You can tell if it's fresh—just look at it and smell it." T. L. watches during his soliloquy and grins. The pair became friends while living in Monroe, Louisiana. T. L., who now lives in New Orleans, is a private chef.

Mike, loquacious and colorful, moved back to Louisiana from Florida to work for the Louisiana Coastal Conservation Association, where he stayed for eight years. "In Florida, you buy bait and come back and cook it," he says with a smirky grin. "In Louisiana, you buy bait, bring back fish and then cook the fish."

They cooked in Mike's log house, set in a beautiful rural area west of St. Gabriel. In the bucolic environment, it is hard to believe that we are only ten miles from Tiger Stadium, ten miles from a Cabela's sporting goods store, and fifty miles from New Orleans. Mike calls it "a pretty good spot." The yard is full of duck decoys, and the inside of the rustic home is completely jammed with outdoors mementos, tanned furs, fish and wildlife mounts, and art. The huge country kitchen is spacious and well used. He makes and jars Big Ike's Barbeque Sauce. "It's my daddy's recipe," he explains. "I just made 140 jars. I cook a lot—five nights a week."

The dish is named after mangrove snappers, one of their favorite fish for the dish, although any other snapper will do as well. T. L. explained that the fish is grilled so that it has a stronger taste than if it is fried. They first used mangoes with it when they were at a camp at Fourchon and liked it.

Tracey Bayles on left and Mike McMullen on right are a team in the kitchen.

Tip: The shallots called for in this dish are true shallots, not green onions.

4 8-oz. snapper fillets
3 tbsp. butter, divided
2 tbsp. minced shallots
½ cup white wine
2 mangos, divided
Salt and pepper to taste
1 tsp. minced fresh basil
1 tsp. minced fresh parsley
Non-stick cooking spray

Wash snapper fillets and set aside. Melt 1 tbsp. butter in a small pan. Add shallots and cook until tender. Add wine and cook until reduced by half. Peel and seed 1 mango and purée in food processor. Add 4 tbsp. of puréed mango to butter mixture, cook until blended, and set aside. Melt 2 tbsp. butter in ovenproof bowl and cook in 350-degree oven until brown. Remove the butter from the oven and brush browned butter onto the fish fillets. Sprinkle with salt, pepper, basil, and parsley. Build a hot charcoal fire in the grill, spray the grates with non-stick cooking spray, and grill the fillets until done, turning only once. Cut 1 mango into wedges and grill alongside the fish. Remove fillets from grill and put on serving platter. Pour the sauce over the fillets. Garnish with the grilled mango wedges. Serves 4.

Polynesian Amberjack

Chris Macaluso of Baton Rouge calls his recipes "off the top of his head—not copied from cookbooks." He likes the cooking channels on TV and loves to talk to chefs. "My love of cooking started as a five-year-old kid when I bugged my mom to learn to make a roux. By six, I was cooking over a stove." He calls himself a journalist who has worked extensively in coastal restoration, but that isn't where his greatest gift lies. It's in cooking. This dish has its roots in a Coastal Conservation Association cooking contest that he was trying to win. He was trying to keep fish moist on the grill, so he wrapped it with applewood bacon, which is mild enough not to overwhelm the fish. The Polynesian part came from his attempt to fuse sweet with spicy. Pineapple provides just enough sweetness and acidity to work well with the aroma and natural heat of jalapeño peppers. The result is a dish that is not too spicy and not too fruity, in spite of the definite presence of fruit. This recipe works well with cobia, and wahoo may also be used if extra care is taken not to overcook this touchy fish.

Fish

1 lb. amberjack loin
1 lb. bacon

Trim off any discolored flesh or fat from the fish. Cut across the loin, making 2-inch medallions. Set aside. Spread the bacon strips on a cutting board and pound them with the back of a cooking spoon to flatten for more coverage and to spread the fat around. Set aside.

Marinade

½ 15-oz. can pineapple chunks
½ cup canned jalapeño peppers, chopped
2 tsp. minced garlic
1 tsp. minced shallot
1 tbsp. minced fresh cilantro
2 tbsp. olive oil
1 6-oz. can pineapple juice

Combine all ingredients and whisk to blend. Pour into zippered

Chris Macaluso is a fanatic for cooking on his Green Egg.

Tip: Soak the toothpicks in water for 1 hour to keep them from burning on the grill.

Tip: Do not use thick-sliced bacon for this dish or it will overpower the fish.

Tip: A spray bottle of water is handy to douse flare-ups, which can char the fish.

plastic bag, add amberjack medallions, and marinate in the refrigerator 6-8 hours.

Salsa

½ fresh pineapple, cleaned and cut into chunks
1 tsp. minced garlic
1 tsp. minced shallots
½ tsp. sea salt
½ cup canned jalapeño slices
Pinch of fresh cilantro

Put all the ingredients in a food processor and pulse until blended.

Assembling the Dish

Creole seasoning, to taste

Lay strips of bacon on a cutting board and cut in half (½ strip for each medallion). Place a medallion in the center of each half-strip. Sprinkle each medallion with Creole seasoning and wrap with the bacon, securing it with a toothpick. Grill on a green egg or charcoal grill at 400 degrees until fish is opaque and the bacon is lightly crisped. Turn each piece to brown both sides. Do not overcook! When almost done, spoon a dollop of salsa onto each medallion and cook for about 2 minutes. Serves 4.

Use toothpicks to firmly secure the bacon to the medallions.

Daigle's Cajun Sweet & Sour Grilled Fish

On a swing through Cajun Country, Gayle Daigle of Church Point, Louisiana, prepared this for us using her Daigle family seasonings and sauces. The Sweet & Sour Sauce is unique—part barbeque sauce, part Chinese sweet and sour sauce. Salt, black pepper, and red pepper can be substituted for her fish seasoning, but some of her special ingredients will be missing. These and all Gayle's ingredients, her other four sauces and her Grillin' Seasoning, can be ordered off the Internet by searching for Daigle Family Company. Gayle's whole family cooks. The sauce recipes are based on recipes from her mother, Cecile Trahan, who at ninety-three still lives independently and cooks. Her dad had a bar and restaurant called Leon's Lounge, where he made the "meanest hamburger around." She says people from all over southwest Louisiana knew about his hamburgers. But she says that she learned the most from her mom. "She taught me how to clean fish and cut up a chicken. She was a good cook!" Although the fish can be cooked directly on a grill, Gayle prefers to use a cast iron griddle placed on the grill. She says that it provides a more controlled heat and it is easier to keep the fillets from breaking up than when they are placed directly on grill grates.

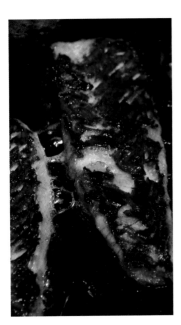

Tip: Use a silicon brush to coat grill with oil because the bristles won't melt.

Tip: Don't flip the fish too early—wait until the fillets are brown. This works best on firm-fleshed fish. Catfish and speckled trout fillets are too soft and break up.

Olive oil
2 lb. sac-a-lait or bass fillets
Daigle's Cajun Fish Seasoning, or salt, black pepper, and red pepper
Daigle's Cajun Sweet & Sour Sauce

Heat grill or griddle to 325 degrees or medium heat on charcoal grill. Pour or brush enough olive oil on griddle or grill to coat the surface and keep the fish from sticking. Season the fillets with Daigle's Cajun Fish Seasoning, or salt, black pepper, and red pepper, and place in a bowl. Pour Sweet & Sour Sauce over fish, mix, and marinate for 30 minutes in the refrigerator. Remove fillets from marinade and grill until the fish is flakey. Brush a coating of Sweet & Sour Sauce on fish. Allow the sauce to caramelize before removing from the grill. Serves 4.

Michael's Speck-Tacular

When Mary Poe, noted fishing guide and co-owner with her husband, Jeff, of Big Lake Guide Service in Calcasieu Parish, called us about this recipe, we got excited. She said, "It is sooo good—it is durn good." Mary, besides being an excellent fisherman, is a great cook and knows good food. She got the recipe from one of her customers, Michael O'Connor of Baton Rouge, Louisiana. "This is quick," she says. "It's healthy too—better than frying."

8 8-10 oz. speckled trout fillets
2 eggs
1 tbsp. water
1 sleeve saltine crackers
¼ cup slivered almonds
1 tbsp. sweet paprika
Creole seasoning to taste
4 tbsp. butter
2 tbsp. olive oil
Juice of 1 lemon

Rinse fillets and set aside. Make an egg wash by beating the eggs with water. Place crackers, almonds, paprika, and Creole seasoning in a food processor and chop. Dredge 4 fillets in the egg wash, then roll them in the cracker-crumb mixture. Melt the butter in a large frying pan and add olive oil. Heat on medium until hot, add fillets, and brown. Turn over and fry the other side. Drizzle with lemon juice. Repeat for second batch. Serves 4.

Tip: Don't crowd the pan or the fish will turn to mush.

Korean-Style Grilled Fish

When we first tried this delightful dish in 2000, we were worried that all the sesame would overpower the fish. We had the option of using vegetable oil, but we went ahead with sesame oil instead. We had sesame oil left over from doing some Asian cooking and wanted to use it before it got rancid. We shouldn't have worried. The sesame accented the fish and the marinade turned the surface of the fish a rich brown color during cooking. This dish is wonderful with any species of snapper, which we normally find a tad dry, as well as catfish, redfish, drum—heck almost anything. Using a folding wire grilling basket makes turning this fish a snap and avoids scuffing the beautiful glaze on the surface of the fish. This one is good for company.

2 lb. fish fillets
2 tbsp. sesame oil
1 cup soy sauce
2 green onions, minced
1 tsp. minced fresh ginger
1 tbsp. minced garlic
1 tsp. black pepper
4 tbsp. sugar
2 tbsp. toasted sesame seeds, crushed

Mix all ingredients except fish and let stand for 1 hour. Put fish and marinade in a plastic bag, mix to coat fish, and marinate in refrigerator for 1-4 hours. Grill over charcoal until fish flakes with a fork (5-10 minutes), turning once. Serves 4 generously.

Fish a la Pepper

This gorgeous dish is beautiful and tastes even better than it looks. The slightly pungent saltiness of the capers is foiled perfectly by the sweet red pepper. Neither overbalances the taste of the fish. Rather, they showcase it. Our cookbook notes tell us that we have been cooking this one since 1994.

1½ lb. fish fillets
2 tsp. chicken bouillon granules
½ cup boiling water
1 tsp. garlic salt
2 tsp. lemon pepper
1 tbsp. olive oil
¼ cup tomato sauce
1 tsp. capers
2 medium red bell pepper, cut into rings

Rinse the fish. Make a broth by dissolving the bouillon granules in boiling water and set aside. Season the fish with garlic salt and lemon pepper. Brown the fish in olive oil in a non-stick pan over moderate heat for 5 minutes, turning only once. Add broth, tomato sauce, and capers to fish. Reduce heat, cover, and simmer 10 minutes. Top with pepper rings and cook an additional 5 minutes or until the fish flakes easily when tested with a fork and the peppers are tender. Serves 4.

Louisiana Gulf Fish Eugenie

Chef Alex Patout invented this wonderful dish when redfish were considered an underutilized commercial species. With wild redfish no longer available on the market, he now most often uses black drum, sheepshead, snapper, amberjack, or cobia. The recipe will work with any firm-fleshed fish, so catfish are out and even speckled trout are marginal. One of the delights of this wonderful dish is how crystal clear the taste of the oregano comes through without overpowering the dish.

Alex, now executive chef at Landry's Restaurant outside of New Iberia, was part of the wave of young chefs who followed Paul Prudhomme in revolutionizing Louisiana restaurant cooking. Prudhomme's K-Paul's Louisiana Kitchen Restaurant opened in New Orleans in July 1979. Patout's Restaurant opened in New Iberia in November 1979. "Paul was gracious enough to send food writers to the country where I cooked," says Alex. In 1988, Alex closed his New Iberia kitchen and opened Alex Patout's Restaurant in the French Quarter. The restaurant was open until 2005, when Hurricane Katrina forced its closure.

Sauce

1 pt. heavy whipping cream
⅔ cup chopped green onions
⅓ cup chopped parsley
1 tbsp. dry sweet basil
1 tbsp. dry oregano
2 lb. peeled crawfish tails

Pour cream into large skillet over medium-high heat. Let the cream simmer, stirring often, and as water evaporates, the cream will thicken. While simmering, add green onions, parsley, basil, and oregano. Test to see if the sauce is thick enough by letting it drip from spoon. When the last drop remains hanging on the spoon, add crawfish. Crawfish will add water and thin out sauce. Continue to stir, bringing cream to a simmer. Continue cooking until desired thickness is once again obtained. While sauce is simmering, begin cooking the fish.

Alex Patout was part of the Cajun Revolution in cooking.

Tip: Alex says be sure to use margarine instead of butter because margarine has a higher smoke point and doesn't burn as easily.

Tip: If fillets from larger fish such as amberjack or cobia are used, the fillets should be re-filleted to no more than ¾-inch thick.

ish

8-10 oz. fish fillets
tsp. salt
tsp. red pepper
2 tsp. black pepper
2 tsp. white pepper
lour
cup margarine

Rinse fillets and pat them dry with paper towels. Season fish by prinkling salt and red, black, and white pepper over both sides of llets. Lightly coat fillets with flour. In a large skillet, melt margarine. t its smoke point, add fillets and cook on medium-high heat, turning ist once. When the fish begin to crisp around the edges, shake skillet o loosen the fillets before you turn them to prevent them from sticking nd breaking. Cook until golden brown. Place on serving plate. Pour generous amount of sauce over each fillet. Serves 6.

Black Drum with Lobster Bisque

Robin Reid became a foodie when she married a Louisiana boy.

Tip: The fish can be cooked in the oven on a rack to allow the drippings to cook off.

Tip: While Robin uses NuNu's Cajun Seasoning for everything, saying that it has something special, she admits that any Creole seasoning can be substituted. Stretch would have her always use his favorite, Chef Leon West's BBQ Seasoning.

Robin Reid learned seafood quick. A native of Tulsa, Oklahoma, she met her future husband, Stretch, in Lafayette sixteen years ago. They promptly got married and moved to rural St. Tammany Parish. She explains, "I only learned to cook when I moved to Louisiana. I never ate white rice till I moved here. When we ate rice, it was flavored from a box. I had to learn. My love of good food grew. I like the praise when I feed somebody. I wasn't a seafood cook until he [Stretch] bought the place." Northshore Seafood, their full-line retail seafood store, became a 2011 victim of reduced demand for seafood after the B.P. oil spill. "I love seafood," she rolls her blue eyes. Stretch concurs. "She could eat fish five days a week." Glenda and I agree that she learned well. The Lobster Bisque recipe is derived from a bisque that she made to sell ready to heat and eat from the store, using lobsters being rotated out of the store's live lobster tank. Robin and Stretch now own and operate In & Out Seafood in Alton, between Slidell and Pearl River, Louisiana.

Lobster Bisque

1 lobster (1 to 1⅛ lb.)
4 tbsp. butter
2 cloves garlic, minced
1 stalk celery, chopped
½ onion, chopped
2 green onions, chopped
1 tbsp. flour
1 cup chicken broth
1 cup half-and-half

In a large pot, bring 2 qt. water to boil. Drop lobster into boiling water, and bring water back to a boil. Turn off fire and let lobster sit in water until cool. Peel the lobster claws and tail. Chop coarsely and set aside. In a separate pot, melt butter. Add garlic, celery, onion, and green onions and sauté until tender. Mix flour with chicken broth and whisk into the sautéed vegetables. Add half-and-half and lobster meat. Set aside, covered to keep warm.

Fish
4 8-oz. drum fillets
NuNu's Cajun Seasoning
Olive oil

Lay the drum fillets on a platter and sprinkle liberally with NuNu's or other Cajun/Creole seasoning. Pour 1 tbsp. olive oil in a frying pan and heat on medium high. Add drum fillets and cook for 2-3 minutes on each side or until brown and the fish flakes easily. Add more olive oil as needed to fry all of the fish. Re-heat the bisque thoroughly and spoon over fish fillets. Serves 4.

CenLA Seafood Jambalaya

CenLA, for the uninitiated, means "central Louisiana." It undisputed capital is Alexandria. This interesting recipe came from Philip Timothy, who was the outdoors editor for the *Alexandria Town Talk* newspaper. We have eaten jambalayas many ways, but this is the only one that we have ever tried that uses a canned soup and the only one that is baked in the oven. It is excellent!

Any size or species of catfish is good in this dish. (Courtesy The Fish Net Co.)

1 lb. small catfish fillets
1 lb. peeled small shrimp
1 lb. crabmeat
3 sticks butter, divided
Creole seasoning
Salt and pepper to taste
Lemon juice
2 cups rice
½ cup chopped onion
½ cup chopped bell pepper
2 cloves garlic, minced
1 10¾ oz. can beef broth
1 10¾ oz. can chicken broth
1 10¾ oz. can French onion soup
½ cup chopped green onions

Cut the catfish fillets into 1-inch pieces. Rinse the catfish, shrimp, and crabmeat and set aside. Melt 1½ sticks of butter in a 4-qt. cast iron Dutch oven. Add fish, shrimp, and crabmeat. Season with Creole seasoning, salt, pepper, and a couple dashes of lemon juice. Saute ingredients until fish and shrimp are done. Remove from pan and set aside. Melt the rest of the butter in the Dutch oven. Add rice, onion, bell pepper, and garlic. Stir mixture continuously over medium to high heat. Do not allow it to stick or burn. When the rice turns brown and crunchy, pour in both cans of broth and the can of soup. Add the fish, shrimp, and crab mixture and stir. Cover and place in a preheated 375-degree oven. Cook for 45 minutes. When done, stir ingredients once, season to taste with salt and pepper, and add green onions. Serves 6.

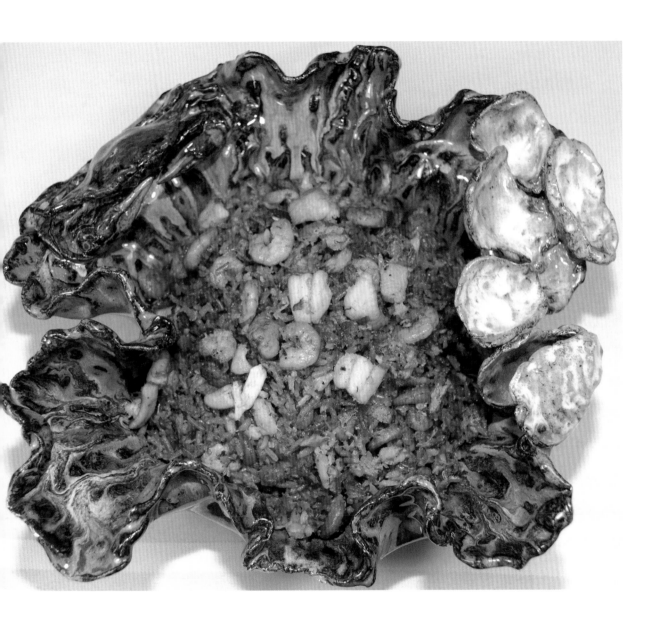

De-Mythifying Cast Iron Cookware

Cast iron remains one of the very best cookware materials available for use today. Properly seasoned, it is non-stick. It browns food wonderfully. It conducts heat well, without having hot spots. It is durable and lasts forever. And food tastes better when cooked in cast iron, probably because a little migration of iron from the pan into the food occurs. So why don't more people use it? Probably because there have come to be so many mythological "don'ts" associated with using cast iron. What follows are a few of them, followed by the facts.

Myth 1. Never wash cast iron or it will rust.

Fact. Dirty pans will make people sick. If the piece is seasoned, it won't rust after washing. Even if it is poorly seasoned, a very light coat of cooking oil rubbed over the interior after washing will prevent rust.

Myth 2. Never use soap to wash cast iron.

Fact. If you aren't using soap to cut the crud, you aren't washing your dishes; you're just rinsing them. It is best not to use strong soap and extremely hot water together, however.

Myth 3. Never cook acidic dishes, such as ones with tomato in them, in cast iron.

Fact. Does anyone really believe that red gravies, sauce piquants, and Creoles were never cooked until Magnalite was invented? Cast iron works wonderfully for these dishes. Just don't store acidic dishes in cast iron or they will discolor the food and export a metallic taste to it. Really, no foods should be stored in any cookware.

Myth 4. Never wipe cast iron dry. You must dry it on a hot stove.

Fact. Drying a piece on a burner is fine. It's also a good way to burn the seasoning off the piece by leaving it on a minute too long. It's kind of hard to dry a Dutch oven lid on a burner too—especially an electric burner. Wiping a piece dry is perfectly acceptable. Dry is dry, no matter how you get it done. Paper towels are better than cloth because cast iron will leave some black discoloration on the wiping material.

Myth 5. Cast iron pieces must be stored with their lids off or they will rust.

Fact. A dry piece, a seasoned piece, or, better yet, a dry, seasoned piece will not readily rust. Moisture will not mysteriously collect and condense inside a piece with the cover on it.

Myth 6. Never boil water in cast iron cookware.

Fact. What did people do before aluminum came along? Were all the cast iron tea kettles built for decoration? Cast iron works wonderfully for cooking greaseless foods like vegetables in water, as well as for boiling just plain water.

Lemon-Peppered Trout

Aulton Cryer Jr. prides himself on the greens that he raises and sells.

Tip: If you use regular lemons instead of Meyer lemons, use 4 lemons.

This recipe comes from Aulton Cryer Jr., a man whose life revolves around food. He's either raising it, selling it, or cooking it. He sells vegetables raised on his Mt. Herman, Louisiana, family farm from his signature white Cryer Family Produce panel truck (an ex-bookmobile) six days a week in Franklinton and Kentwood, Louisiana, and Tylertown, McComb, and Magnolia, Mississippi. He says that his produce-vending philosophy is based on healthier living. "I eat what I sell; a diet based around vegetables that will unclog your arteries. I add a good bit of seafood and an occasional steak and some poultry now and then." He inherited the commercial vegetable farming bug from his father, Aulton Sr., who he says was famous for his greens—turnips, collards, and mustard—and, of course, Washington Parish watermelons. Aulton, a graduate of the Epicurean School of Culinary Arts in Hollywood, loves to cook. "I cook because I love it. If there was one thing that I could do the rest of my life, it would be cooking. It speaks to my creative side."

4 8-oz. trout fillets
2 Louisiana Meyer lemons, juiced
1 tbsp. sea salt
1 tsp. black pepper
1 tbsp. minced onions
2 tbsp. garlic powder
½ cup butter
Cooked rice

Layer the fillets in the bottom of a shallow bowl. Pour the lemon juice over fillets. Marinate for 30 minutes. In another shallow bowl add the salt, pepper, minced onions, and garlic powder and mix well. Roll the fillets in the seasonings. Melt the butter in a large pan and sauté the fish in it for 1-2 minutes on each side. Add the lemon juice from the marinade over the fish, cover, and simmer for 2 additional minutes or until the fish flakes easily when tested with a fork. Drizzle the lemon sauce over rice and serve with the fish. Serves 4.

Blackened Fish Tacos

Louisiana food and fishing brought Mark Vicknair back to Louisiana from California.

Tip: Be careful when turning the fillets because they easily stick to the pan.

Mark Vicknair of Laplace, Louisiana, who prepared this recipe for us, was always the kid who used to run the Mississippi River levees on his bike. He would fish in the ponds behind the levee and the river itself for catfish and perch. With a huge grin smeared over his face, the fifty-three-year-old admits that at ten or eleven years old he would sneak into the big river for swimming. "The way we got caught was our underwear would get brown from the river water. We would get a spanking. We got smart after getting caught. We would skinny dip or we would leave the brown drawers behind, hanging in a tree behind the levee to use the next time. Mom would say, 'You been swimming in the river?' I'd say, 'Nope, not me.' I used to do things that I would never think about having my kids do now."

These days, he is a speckled trout fisherman who fishes in Venice, Grand Isle, Cocodrie, and even Cat Island in Mississippi. "The first time I made blackened fish tacos was at a friend's camp in Cocodrie. I had eaten it in a restaurant and said, 'Man, this is good! I got to try this.' I didn't tell the other guys at the camp what I was going to make. They started frying fish and I started blackening for tacos. After they tried one, they wanted more. It shut down the frying."

Salsa
3 large tomatoes, or 5 cluster tomatoes
3 jalapeño peppers
3 yellow sweet banana peppers
1 medium yellow onion, chopped
3 green onions, chopped
2 tsp. chopped cilantro
1 tsp. salt
1 tsp. garlic powder
Juice of 1 lime

Boil tomatoes and peppers in water until tender. Put chopped onions, green onions, and cilantro in a bowl. When tomatoes and peppers are done, remove and discard pepper stems and tomato skins. Put tomatoes and peppers in a blender and lightly purée. Pour over chopped ingredients and mix well. Add salt and garlic powder

Add lime juice and let cool. This may be made ahead and stored in a refrigerator for 7 to 10 days.

Slaw

2 tbsp. Cardini's Caesar dressing
1½ cups finely shredded cabbage

Pour the Caesar dressing over the cabbage and mix well. Set aside.

Guacamole

3 small avocados
1 tbsp. sour cream
¼ tsp. salt
¼ tsp. garlic powder
1 tbsp. Salsa
1 tbsp. minced onion

Cut the avocados in half and remove the seeds. Scoop the avocado flesh from the peel with a spoon. Mash the avocado with a fork, add the other ingredients, and mash again. Stir to mix well.

Finishing the Tacos

1½ lb. small to medium fish fillets
Chef Paul Prudhomme's Blackened Redfish Magic
2 tbsp. chili powder
Pinch of sugar
2 tbsp. butter
8 flour tortillas
1 cup shredded Mexican Four Cheese Blend

Rinse fillets, pat dry, and season generously with Blackened Redfish Magic, chili powder, and sugar. Melt butter in cast iron skillet over high heat. Place fish seasoned side down in skillet, then season the other side. Turn fillets only once, after 3-5 minutes, when fish are dark brown, nearly black. Cook the second side 1-2 minutes depending on thickness of fillets. Warm the tortillas on a griddle or in a pan. Lay a piece of fish on center of each tortilla. Layer guacamole, slaw, salsa, and cheese on the fish. Fold tortillas over and serve. Serves 4.

Deano's Dago
Blackened Redfish

"I tried some of the other recipes for blackened fish and I wasn't that crazy about them," says Deano Bonano of Metairie, Louisiana, "so I put this together. A lot of places blacken fish, but they put so much gook on it that it is all you taste. The Italian dressing on these gives a nice little tang." Like other true blackening preparations, it is almost necessary to do it outside on an outdoors burner. It generates copious quantities of smoke that can overwhelm a kitchen.

Deano named his dish lightheartedly because he is proud of his Italian heritage. "My paternal great-grandparents were from Italy. They had a grocery store in the old Carrollton neighborhood of New Orleans—you know, the old corner grocery stores that are now extinct. My grandmother Rosalie Bonano would start cooking her red gravy for Sunday on a Saturday afternoon. Except during Lent, it had everything in it: beef roast, pickled meat, meatballs, and sometimes ham hocks. During Lent, it just had eggs. She taught me to fry down the tomatoes and tomato paste until it was brown to give the gravy that smooth taste."

8 medium redfish fillets
Ken's Steakhouse Light Northern Italian Dressing
1 cup melted butter, divided
2 tbsp. lemon juice
1 tbsp. minced garlic
2 tbsp. Italian seasoning, divided
1 tbsp. garlic salt
1 tbsp. onion salt
2 tbsp. Creole seasoning
Salt and black pepper to taste

Place the fillets in a bowl. Pour enough Ken's Italian dressing to cover the fillets and marinate in the refrigerator for 30 minutes. In a small pan, add ¾ cup of butter, the lemon juice, minced garlic, and 1 tbsp. Italian seasoning and sauté to blend flavors. Strain the butter mixture and retain the liquid. Set aside. In a small bowl, combine garlic salt, onion salt, 1 tbsp. Italian seasoning, Creole seasoning, salt, and black pepper. Remove fillets from marinade and sprinkle

Deano Bonano blackens his fish outside because of the quantities of smoke produced by the technique.

Tip: Avoid putting too much garlic butter in the pan because the fish won't blacken properly.

Tip: Remove the red blood line from the skin side of the fillets. Larger fish may be cut into smaller serving size pieces.

both sides of the fish with the dry seasoning mix. Place an iron skillet over a very hot fire and heat until a drop of butter when dropped in pan begins to immediately smoke. Pour enough garlic butter into the pan to coat the bottom thoroughly. Allow the butter to heat up for 30 seconds, then add fish fillets, making sure the fillets do not touch each other. Cook the fillets for several minutes on each side, until they are crispy dark brown, but not black. Add ¼ cup butter to garlic butter and heat. Pour the garlic butter over the top of the cooked redfish fillets and serve piping hot. Serves 4.

Cashew-Crusted Fish

Duane Taylor fishes hard when the time is right to fill his freezer with ops, his favorite eating fish.

Tip: Be sure to dredge the fish in the crusting mixture in small batches so that the excess butter doesn't make the mixture clump up into an unusable mass.

Tip: A food processor is ideal for crumbling the cashews called for in this dish.

Tip: It takes a lot of heat to make the cashew crust crispy. If the test piece placed on the griddle doesn't crisp properly, increase the temperature of the griddle.

Duane Taylor of West Monroe spends almost every moment that he isn't asleep in the outdoors. His job as a game warden with the Louisiana Department of Wildlife and Fisheries keeps him outside, and if that isn't enough, he spends most of his free time fishing and hunting. This is a Duane Taylor original and his favorite way to cook fish, especially with his favorite fish, yellow or Opelousas catfish, which he invariably calls "ops." He also has a favorite part of the fish to cook, cross-sections of the thick belly flaps of the fish. Ops bellies have a mild taste and an almost buttery texture that is hard to duplicate with another fish. The dish will work with any fish, however. It is best done with a griddle with an accurate temperature control, although a heavy skillet can be used.

3 lb. fish fillets cut into strips ½-inch thick
2 cups coarsely crumbled cashew halves and pieces
1 cup all-purpose flour
1 tbsp. onion powder
1 tbsp. garlic powder
½ tsp. red pepper
1 tbsp. salt
1 tbsp. black pepper
¼ lb. butter, melted

Rinse the fillets and pat dry with paper towels. Grind the cashews with short pulses in a food processor until they are the consistency of cornmeal. Do not overprocess or they will turn into something resembling cashew butter. Toast the ground cashews on a baking sheet in the oven for 20 minutes at 200 degrees to dry them slightly. Do not brown them. Mix the cashews, flour, onion powder, garlic powder, and red pepper. Add salt and black pepper. Melt the butter then dip the fish in the butter to coat the fillets. Spoon a modest amount of the cashew breading into a working bowl so that the butter from the first fish does not make all the breading soggy. Roll the fish strips in the breading one at a time to thoroughly coat them. Place them on a heavy skillet or griddle that has been lightly coated with butter and heated to very hot but before the butter smokes. Repeat until all the fish are used.

Cook about 2 minutes or until a golden crust forms on the breading beneath the fish and the flesh becomes opaque. Carefully turn the fish with a spatula to cook the other side. Serves 6.

A professional griddle produces wonderful results.

Honey-Pecan Trout

Brandi Masson of Metairie, Louisiana produced this masterpiece. She got the idea from a Chinese restaurant that had honey-pecan shrimp on the menu. One day, she planned fried fish at home and she said to herself, "Let me see if I can make that sauce." She did, and it's good. She says that she ate a lot of seafood growing up and still does. Husband Todd is an avid fisherman, besides being the publisher of laspecks.com. Brandi, an accountant, says she fished a lot when they first got married and she still loves to crab with her family. "We try to eat healthy," she says. "Seafood is healthy, but mostly we eat it because we love it."

Son Joel Masson actively cooks with his parents, Brandi and Todd.

Tip: Don't batter the fillets too far ahead or the breading will fall off.

Tip: When frying the fish, fry the flat (skin) side of the fillet down first to keep the ends from curling up.

8 medium trout fillets
1 cup pecans, divided
3 tbsp. honey, divided
1 cup panko breadcrumbs
1 cup French bread crumbs
1 cup milk
1 cup flour
1 tsp. Creole seasoning
1 tsp. salt
Butter
Olive oil
1 cup whipping cream
1 tbsp. Dijon mustard

Rinse fillets, pat dry with paper towels, and set aside. Spread ¼ cup of pecan halves on a baking pan. Drizzle 1 tbsp. honey over pecans and toast in a 350-degree oven until pecans are glazed and toasted, about 3-5 minutes. Chop remaining ¾ cup of pecans in a food processor. In a large bowl, combine panko breadcrumbs with French bread crumbs and chopped pecans. Put milk in a separate bowl, and in a third bowl, mix flour with Creole seasoning and salt. Roll the fillets in the flour mixture, dip them in milk, and then roll them in the pecan and breadcrumb mixture. Heat 2 tbsp. butter and 2 tbsp. olive oil in a large skillet over medium-high heat. Add fillets a few at a time and fry until golden brown and they flake easily with a fork

Add more olive oil and butter as needed to cook the fish. In a separate saucepan, over medium heat, reduce the whipping cream until thick. Add Dijon mustard and 2 tbsp. honey. Spoon cream sauce over fillets and sprinkle with toasted pecans. Serves 4-6.

Bronzed Catfish

Gary Hayes is a cooking junkie. "A phenomenal cook" is what wife Adele calls him. His Kenner neighbors are always dropping by to see what he has cooking. "I think that I am officially kicked out of the kitchen," she adds. When Adele met Gary, she had "tons" of Paul Prudhomme's seasonings, so he began to experiment with different dishes. Adele doesn't eat much fried food, so Gary started bronzing fish. Bronzing instead of blackening is easier, he explains. "It's not as spicy and you don't have to blacken outside the house. Blackening is good; it's fine; it's crusty. But when you come home from work you can easily do this." In his version of bronzing, he uses olive oil in a non-stick skillet instead of butter. Gary catches his own eight- to fourteen-inch catfish from Lake Cataouache.

1½ lb. 5-7-inch catfish fillets
Olive oil
Paul Prudhomme's Seafood Magic

Wash fish thoroughly and pat dry with paper towels. Rub olive oil on fish fillets and generously sprinkle Paul Prudhomme's Seafood Magic on one side of each fillet. Rub the seasoning in to make sure it sticks. Add 1 tbsp. olive oil to a non-stick skillet and heat on medium-high heat. When the pan is hot, lay the fillets in it with the seasoned sides down. While they are cooking, sprinkle Seafood Magic on the unseasoned sides, which are facing up. Cook 3 minutes, then turn the fillets over. Cook an additional 3 minutes. Serves 4-6.

Gary Hayes knows his fish is fresh because he catches it in Lake Cataouache.

Tip: Be sure the pan is hot before adding the fish to keep it from getting soggy.

Tip: Cook the skin side down first to keep the fillets from curling.

Shuck-a-Fied Snapper

Bert Istre on left and David Bertrand are friends as well as partners in the restaurant.

Tip: Resist the temptation to flip the fish too often. Flip it once when you can feel a crust developing on the first side.

Tip: This dish can be made with any white-fleshed fish, including catfish. But if catfish are used, cook them skin side down first to reduce curling of the fillets.

A more open-armed Cajun welcome could not have been provided to us than what David Bertrand and Bert Istre gave when we visited Shucks! Restaurant in Abbeville, Louisiana. The dynamic and good natured pair plied us with some of their new creations—like Shuck-a-Fied Snapper—and a whole lot of their menu traditions. This scrumptious dish was invented by David and approved by Bert, who said, "When I first tried it, I was lovin' it!" They have done a lot with the restaurant, but then again they had a lot to start with. It was first opened as Dupuy's Oyster Bar in the 1800s before being purchased in 1973 by Harold Hebert, an oyster fisherman, and his wife Doris, originally from Golden Meadow, Louisiana. A scant six successful years later, their daughters, Linda and Diane, bought the place. Diane and her husband, Jack Phares, operated it until 2007. Midway through, in 1995, they left their old quarters for a larger, more modern building and changed the name to Shucks!

In 2007, both David and Bert were negotiating independently to purchase the business. They knew each other, and in fact, Bert had done an internship at David's restaurant, Bertrand's Riverside. They paired up to make what David calls "a match made in heaven." Bert points a finger at his partner, "He's the one who loves to play with the food." David replies to Bert, who holds a degree in restaurant management, "I don't want to know about the business side of it. It messes with my creativity."

Au Gratin Sauce

1 stick butter
1 large onion, chopped
½ cup flour
2 pt. half-and-half
1 tbsp. chicken base
1 tbsp. granulated onion
1 tsp. granulated garlic
1 tsp. red pepper
¼ cup grated Romano/Parmesan cheese
½ cup shredded Parmesan cheese
1 cup American cheese

Melt butter in a large pan. Add onion and sauté until tender. Stir in flour and mix well. Pour in half-and-half and chicken base and stir until mixed. Add remaining ingredients and simmer over low fire until the sauce thickens. Cornstarch may be added to thicken sauce if necessary.

Roasted Red Bell Pepper Sauce

½ roasted red bell pepper
2 pt. heavy cream
1¼ tsp. granulated onion
1¼ tsp. granulated garlic
1¼ tsp. Tony Chachere's Original Creole Seasoning
½ tsp. red pepper
1¼ tsp. dried dill weed

Purée the bell pepper in a food processor. Add all ingredients to a pan and simmer until thickened.

Fish

4 6-8 oz. snapper fillets
2 tbsp. butter
1 lb. lump or jumbo lump crabmeat

Red snapper is the fish Shucks! uses for this dish.

Melt butter on a griddle or pan and sear the fillets on both sides until the fish flakes easily when tested with a fork. Mix 2 parts Roasted Red Bell Pepper Sauce to 1 part Au Gratin Sauce and heat until blended. Add crabmeat and fold in gently. Spoon the sauce over the finished fillets and serve. Any remaining sauce may be refrigerated for later use. Serves 4.

Pan-Fried Seasoned Fish

Grouper are one of Carlos Valdez's favorite fish.

Tip: Be sure not to overcook the fish and turn it only once. Repeated turning or overcooking will result in the fish falling apart.

Commercial fishermen spend a lot of time on their boats, so many become skilled cooks. Not just Louisiana fishermen cook, though. This recipe comes from Carlos Valdez, a Florida grouper and stone crab fisherman. Naturally, since he is a grouper fisherman, he uses grouper, but we have used snapper, redfish, drum, bass, and other white-fleshed fish and they work just fine. The cumin in the recipe gives away its Latin roots. Don't skip it—it won't be the same. The paprika is mainly for color, but the soy sauce is critical, since it is the only source of salt in the recipe. Carlos and his wife, Dulcie, have opened a smash hit restaurant, Havana Café, in Chokoloskee, Florida, on the edge of Everglades National Park. This recipe, as well as others from his fishing boat, are on the menu there.

1½ lb. fish fillets
⅓ cup soy sauce
1½ tsp. cumin
4 large cloves garlic, minced or pressed
½ tsp. black pepper
½ tsp. paprika
⅓ cup olive oil
2 fresh limes

Cut the fillets into serving-size pieces and set aside. Combine soy sauce, cumin, garlic, pepper, and paprika in a small bowl. Spoon mixture over the fish fillets and rub it in. Add olive oil to pan and when hot, sear the fillets on both sides until the flesh is opaque, about 5 minutes per side. Remove from the pan and squeeze fresh lime juice over the fillets. Serves 4.

Escabeche

This classic Spanish dish is ideal for oily fish such as mackerel, bluefish, and jacks of all kinds. It also works well with lean, white-fleshed fish, but there are a zillion ways to prepare them. In Spain, even herring is used, a fish we consider too boney to eat. The result is tangy and light, perfect for a summertime luncheon. It is almost like a fish salad, but isn't. It is well worth adding to your repertoire of recipes. We liked it the first time we tried it.

1 lb. fish fillets
¼ cup kosher salt
4 cups water
¼ cup olive oil
2 cloves garlic, mashed
1 jalapeño pepper, cut in half
4 bay leaves, divided
1 large onion, sliced in half-moons
1 tsp. black peppercorns
½ tsp. cumin seeds
1 tsp. dried thyme
1 tsp. coriander seeds
1 tbsp. dried oregano
1 cup fish or chicken broth
1 cup white wine
1 cup white wine vinegar

Amberjack are one of several excellent choices for this recipe.

Cut fish into 2-3 inch pieces and set aside. Combine salt with 4 cups water, stir to mix, and soak the fish in the brine for 30-45 minutes. Heat the olive oil, garlic, jalapeño pepper, and 2 bay leaves in a large pan over medium heat until the garlic browns, about 4-6 minutes. Do not let the garlic burn. Remove the garlic, jalapeño, and bay leaves and discard. Turn the heat up to medium high and add the fish. Sear on both sides for 1-3 minutes, depending on thickness of pieces. The fish is best not cooked all the way through. Remove the fish to cool and then add the sliced onion, lower the heat to medium and cook until translucent. Remove the onions to cool. Add the remaining ingredients, turn the heat to high, and bring to a rolling boil. Reduce y half. Turn off the heat and let cool. When all ingredients are at oom temperature, pour the sauce into a container, and add the fish nd onions. Store in the refrigerator overnight to let the flavors marry. erve cool or at room temperature. Serves 4.

Leanna's Chinese Catfish

The claim to fame of Frank's Lounge in Des Allemands, Louisiana, was their Bloody Marys. But to me, this is much better. In 1984 Frank's wife, Leanna Kraemer, whipped up some of these behind the bar, and I have loved the recipe ever since. It has a distinct Chinese stir-fry flavor, even though the heat used is lower than in true stir frying and butter is used instead of oil. Using a wok is okay for this but the fillets will sear better in a flat pan.

1 lb. very small catfish fillets
Creole seasoning
6 tbsp. butter, divided
Soy sauce
1 medium bell pepper, slivered
3 medium onions, slivered
2 carrots, bias cut
2 stalks celery, bias cut
Cayenne pepper

Season fillets generously with Creole seasoning. Melt 3 tbsp. butter in a pan over low heat. Spread the fillets skin side down first in a single layer in the pan and sear both sides over medium-high heat. Remove fish, sprinkle heavily with soy sauce and set aside. Melt the rest of the butter and add vegetables. Add soy sauce and cayenne to taste and stir-fry about 1½ minutes over medium-high heat until vegetables are tender but still firm, turning and tossing constantly. Return fillets to pan and very gently stir into the vegetables. Cover and reduce heat to low. Heat until fish are hot, when flesh inside the fillet is flakey and white, absolutely no more than 5 minutes. Serve immediately over rice. Serves 4.

Freshwater catfish, the smaller the better, are what this dish is about.

Tip: Use the tiniest catfish fillets that you can find or catch. You definitely don't want thick slabs of fish for this dish.

Tuna Anthony

Anthony Puglia cooked this dish for us after a successful tuna chasing trip out of Venice. "Whenever I have company over, it' because I just caught a bunch of fish or killed a bunch of ducks," h says while cooking. "I'm not going to have people over for steaks I want to cook something for them that they don't get to eat a lot. Anthony says that he ate tuna a lot growing up. His parents both cooked it, and customers at their family business, Puglia's Sporting Goods in Metairie, would bring tuna steaks in to them. Anthony called the dish "experimental" but to us it was a polished product Combining the tuna with the cheesed spinach and mangoes produce a synergy of tastes—the sea, salty, herby, and sweet.

Anthony Puglia with a yellowfin tuna destined for his kitchen.

Tip: Like any experienced tuna chef, Anthony preaches against the danger of overcooking the fish. Overcooking is easy to do and will completely ruin the dish. The tuna must still be pinkish-red inside to be at its best.

Tip: Anthony recommends using Kosher salt rather than table salt. He says that it has a distinct taste that works well in this dish.

4 6-oz. tuna steaks, 1½-2 inches thick
¾ cup canola oil
1 egg
1 cup milk
1 cup flour
1 cup Italian breadcrumbs
Kosher salt to taste
Black pepper to taste
2 tsp. olive oil
1 clove garlic, sliced thin
10 oz. fresh baby spinach, stems removed
1 cup freshly grated Romano cheese
1 pt. refrigerated mango wedges

Trim any dark red meat from the edges of the tuna. In a medium skillet, heat oil to 350 degrees. Make an egg wash by beating the egg with the milk. Put flour in a shallow bowl and roll the tuna steaks i flour to coat all sides. Dip the steaks in the egg wash and dredge ther in the breadcrumbs. Season both sides of the steaks with salt and pepper. Place tuna in the preheated oil and cook on high until golde brown, approximately 2 minutes on each side. The top and bottom c each steak should be lightly browned. In a separate pan, add 2 tsp olive oil and garlic. Sauté until garlic is tender. Add spinach, salt and pepper to taste, and stir until spinach is wilted. Remove from hea

nd add Romano cheese. When cheese is melted, remove spinach
rom pan. Serve with tuna steaks and slices of mango. Serves 4.

Deep-Fried Stuffed Flounder

A good variety of fish is hard to find in New Orleans, hard as that is to believe. Good shrimp, crabs, crawfish, and oysters are easy to find. Wild Louisiana fish—not so easy! Economic and regulatory restraints have taken their toll. Sadly, the finfish line-up for the majority of Louisiana retail seafood markets is farm-raised catfish (almost always from out of state), farm-raised tilapia (from out of the country), and farm-raised salmon (either from out of state or out of the country).

Fisherman's Cove Seafood Market on the corner of Williams Boulevard and Thirty-Second Street in Kenner is a major exception. When I tell the owner, David Robinson, so, he answers in his clipped business-like way, "I think we could be better." He goes on to explain that the success of the thirty-two-year-old market and its twenty-three-year-old companion restaurant, Harbor Seafood and Oyster Bar, is due to having developed the right suppliers for what they want to sell. "We only sell Gulf seafood—no imports," he stresses. This includes uncommon items such as cobia, mahi, yellowfin tuna, and amberjack, as well as gems such as bluefin tuna and Louisiana speckled trout, a rarity since the gill net ban. Of course, they have flounder, which they do a remarkable thing with. They stuff it (lots of people do that) and then they deep fry it (not so many people do that). Their double-battering technique seals in flavor and moisture like no other way I have eaten flounder.

Preparing the Fish
3 1½ lb. flounders

Scale the flounders, remove their heads, and gut them. Wash the fish well to remove loose scales. Make a pocket by slitting the flounders down the center of the dark side. Insert a knife between the flesh and the bones and loosen the bones from the flesh toward both outside edges. Set aside.

Cutting the pocket in the flounder to stuff it is easier than it looks.

David Robinson lives on the telephone in the search for fresh fish.

Tip: This is Harbor Seafood and Oyster Bar's standard recipe for making crabmeat stuffing. It will make more stuffing than needed for 3 fish. Either reduce the quantities to one-fourth of those given or, better yet, make stuffed crabs with the remainder, as they do for the store and restaurant.

Making the Stuffing

- loaves French bread
- cups chopped onions
- cup chopped bell pepper
- cup chopped celery
- bunch green onion, chopped
- tbsp. chopped parsley
- tsp. salt
- tsp. pepper
- cup canola oil
- lb. lump crabmeat
- lb. crab claw meat

Dipping the stuffed flounder in eggwash twice and then dredging it twice, once in flour and once in corn flour, provides a perfect crust.

Toast French bread in oven or use stale French bread. Break up the bread and add water until moist. Add all seasonings and oil and mix together. Fold in the lump crabmeat and crab claw meat. Spread the mixture on a baking pan and bake in a 350-degree oven for 30-5 minutes. Stir every 10 minutes to prevent sticking. Remove from oven to cool. Place a generous ball of stuffing in the pocket of the flounders so that it is mounded out of the pocket. Press the pocket flap edges over the stuffing.

Finishing the Dish

- cups milk
- eggs
- cups flour
- cups corn flour

Make an egg wash by whisking together the milk and eggs. Place the flour and corn flour in separate bowls. Dredge the stuffed flounders in the egg wash, then the flour, then back in the egg wash, and then in the corn flour. Deep fry at 350 to 375 degrees until golden brown. Serves 6.

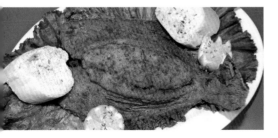

The Secret to Frying Fish

Fish can be cooked a lot of ways—and they all taste good. But catch most Louisianians in a candid moment and they will admit that there just isn't anything like really good fried fish. Fish and frying just seem to go together. But a lot of people make a mess of it at home. The fish is often soggy with cooking oil. And frying makes a mess when done indoors. Oil spatters are everywhere, and it leaves a lingering smell that even fried fish lovers don't love.

We tried everything—pan frying and deep-frying in a Dutch oven on the stove, electric counter-top basket fryers, and deep-frying outdoors over a boiling rig. The electric basket fryers were very easy to overload, so their oil temperature would plummet when too much seafood was added. Outdoor pot frying was a little better because larger oil volume meant that it wasn't as easy to overload as the smaller fryers were. But with both of them, we still had to cool the oil and funnel it into a storage container. Then we had a messy fryer to clean.

Finally Janice Clark, who with her brother, Roger Pitman, owns Pitman Metal Works in Dixie Inn, Louisiana, near Ruston, wired us up to get one of their propane-powered Pitman Fryers. It was the double-barreled version that held two baskets, and it was the best thing (in cooking) that we ever did! That's saying a lot because we own a huge array of cooking equipment.

The large volume of oil (five gallons) means that it is impossible to overload and every piece of food comes out perfectly sealed and crispy on the outside. The volume of oil concerned us at first, because of its expense. No more!

When we used other equipment and drained the pan after every use, we were used to discarding the oil after two fryings. The oil in our Pitman Fryer lasts twelve to fourteen months, without ever being removed from the fryer, and it still produces perfectly fried food. Part of the secret may be that stray breading particles that fall off the food, sink into the deep-V bottom, below the source of heat, instead of setting on the bottom of a pan just above the direct heat.

Its only drawback is also part of its strength. The intense heat provided by the propane very quickly brings oil temperature up to 350 degrees and, in the blink of an eye, to over 375 degrees. Users quickly learn to watch the fryer closely and throttle back the heat when it nears 350 degrees. The temperature will not break over 5 degrees downward when the seafood is added.

Pitman Fryers may be purchased at a few retail outlets, but because the company does not sell at wholesale prices, few stores carry them. They are more easily obtained directly from the factory through the Pitman Fryers website.

Pitman Fryers have the capacity to turn out large volumes of perfectly fried food, but also work superbly for making dinner for one.

Buffalo Patties

Reggie Craig grinds the buffalo flesh and the small bones disappear.

Tip: Be sure to even out the edges of the patties before frying. If they are too thin, they will break off during frying.

Reggie Craig of Deville, Louisiana has been around fishing all his life. His father, Fred, was a freshwater commercial fishermen plying his trade with hoop nets, trot lines, gill nets, trammel nets and seines. His targets were mainly catfish, gaspergou, and buffalo. Reggie himself is the plant manager for Catahoula Manufacturing in Jonesville. They make netting, twines, and ropes for commercial fishermen. "Daddy had a talent for it," he says. "He knew where the fish were. Some people think that all they have to do is buy some nets to be a fisherman, but it's a lot harder to make a living at it than people think." Fried buffalo ribs have always been a delicacy for those familiar with them. The bony part of the buffalo fish—which is most of it—is difficult to eat, being filled with tiny hair-like bones floating in the flesh. This recipe uses that meat, wonderfully white and flakey, and the bones disappear.

Reggie credits this recipe to his grandfather, Loyd Mabou. Everybody wanted the ribs, his grandfather said, and they had to do something with the bony part of the fish. This was it. "He was a pretty good cook. He told me, 'canned and fancy fish is what them dusk-to-dawn light people [city folks with street lights] ate.' He was talking about canned mackerel and salmon in grocery stores. Nothing is fancy here. I don't have any delusions about being a chef," he twanged in his country accent. But he knew it was going to be good. And it was. You will need a meat grinder for this recipe.

5½ lb. buffalo fillets
2 medium onions, chopped
2 slices bread
2 cups instant mashed potatoes
2 tbsp. salt, divided
2½ tsp. black pepper, divided
2 tsp. Creole seasoning, divided
1 tsp. cayenne pepper, divided
2 cups flour
Oil for frying

Cut the fillets into pieces that fit in a meat grinder. Grind the fillets

hen grind the onions into the fish. Put bread through grinder to
push all the onion out. Once all of the onion has been added the fish
discard the bread. Mix the mashed potatoes into the fish and onions
by kneading it. Add 1 tbsp. salt, 1½ tsp. black pepper, 1 tsp. Creole
seasoning, and ½ tsp. cayenne pepper and mix well. Form the mixture
into balls about 3 inches in diameter, then flatten them into patties.
In a separate bowl, mix flour with the remainder of the seasonings.
Dredge the patties in the flour mixture and fry at 350 degrees for
approximately 7 minutes. Flip the patties and fry an additional 3-5
minutes until golden brown. Serves 8-10.

Fried Fish with Capt'n Jack's Cream Sauce

We met Jack Oser and his wife, Janis, at a Westwego Farmers and Fisheries Market concert, where, togged in his chef's whites, he was dispensing his goodies to music and food lovers. It had a laid-back atmosphere and locals came and went all evening. From Algiers, the pair own and operate Capt'n Jack's Gourmet Meals, a catering company that specializes in smoked foods and Cajun-Creole dishes. Jack, a third-generation German from Algiers, the Westbank suburb of New Orleans, and Janis, who split time growing up between Bridge City and New Orleans' Irish Channel, met when she was his boss at West Jefferson Medical Center. Both were X-ray technicians. "We shared a love of food," she said, smiling shyly. "On a cold rainy day we would start cooking. We cooked for the whole week coming. It's nothing for us to cook a forty-quart pot of gumbo. We cooked together for six years and were best friends. I didn't want to kiss him at first. I thought, 'Oh gosh! This is going to mess things up.'"

It hasn't.

Jack and Janis Oser share a love of cooking.

Tip: Jack says you can add shrimp, crawfish tails, or lump crabmeat to the sauce to add more flavor. The sauce may also be served over pasta, steamed vegetables, or fried soft-shell crabs.

Fried Fish

4 8-10-oz. fish fillets
1 egg
¾ cup milk
1 cup seasoned fish fry
Oil for frying

Wash fillets and set aside. Mix egg with milk and beat with fork to mix. Soak fillets in the egg and milk wash for 5-10 minutes. Remove fillets and dredge in fish fry. Fry fish at 350 degrees until golden brown. Remove from pan and drain on paper towels. Set aside in a warm place.

Cream Sauce

stick butter
medium onions, chopped
cup minced celery
bunch green onions, chopped
cloves garlic, minced
cup flour
qt. whole milk
pt. heavy cream
cup chopped parsley
tbsp. Worcestershire sauce
tsp. salt
tsp. white pepper
tsp. cayenne pepper
tsp. hot sauce

Melt the butter in a large saucepan. Add onions, celery, and green onions and sauté until tender. Add garlic and cook until softened. Stir in the flour and whisk until the flour has completely blended with the butter sauce. Add half of the milk and half of the heavy cream and mix well. Add more milk and cream as needed to achieve the desired thickness of the sauce. When thickened, add the remaining ingredients and cook an additional 5 minutes to blend the flavors. Spoon the sauce over the fried fish fillets and serve. Serves 4.

Fish and Crabs on the Ridge

Melanie Charpentier belongs to a Lafourche Parish cooking club that meets and cooks once a month.

Tip: Melanie says that she prefers speckled trout for this dish but often uses red snapper cut into strips as well.

Melanie Charpentier, who invented this dish, was born and raise in Golden Meadow, Louisiana, although she splits time now betwee homes in Lockport and River Ridge, Louisiana. Her husband, Rober was a commercial shrimper until 2005, when he sold his offshor shrimp boat. Melanie maintains that Robert is a good cook too, wit one of his specialties being jambalaya. "I'll never forget the first tim that I invited Robert to eat. I didn't have any shrimp for the jambalaya so I cooked it with just smoked sausage. He kept saying somethin was missing, but he wouldn't come out and say what." Melanie is serious cook and a member of a Lafourche Parish cooking club tha meets to cook once a month. She used to own the Cajun Chill an Grill, a Lockport restaurant. She closed it after a year and a half. "I was like a prison," she explains. This dish showcases the tastes of th fish and the crabmeat.

½ cup breadcrumbs
1 egg, beaten
3 tbsp. mayonnaise
2 tbsp. heavy cream
½ cup finely chopped green onions
2 tbsp. Worcestershire sauce
1 tsp. garlic, minced
Salt and pepper to taste
1 lb. lump crabmeat
2 lb. trout fillets
½ tsp. Morton's Nature's Seasonings Blend
1 10-oz. box seasoned fish fry
Oil for frying
1 cup heavy cream

In a bowl, mix breadcrumbs, egg, mayonnaise, heavy cream, gree onions, Worcestershire sauce, garlic, salt, and pepper. Add crabmea and mix thoroughly. Form into balls about 1½ inches in diamete Season fish with seasonings blend. Wrap a fillet around each crab bal and secure with a toothpick. Roll in seasoned fish fry and drop int hot oil. Fry until golden brown. Simmer cream to reduce its volum

o half. Pour over fish/crab roll. Great as an appetizer or main dish
erved with pasta. Serves 4-6.

Fried Garfish

This recipe belongs to Rusty Munster of Slidell. An avid saltwater sport fisherman, Rusty has been eating gar for twenty-two years. "It's a meat that I never expected to taste good but does." He got started through friends from Marksville, Louisiana. "They serve it there a lot," he says, "like you and I have meatballs and spaghetti." This fish has a mild taste and a dense, firm texture, almost like lean pork.

2 lb. fresh gar fillets
2 eggs
1 cup milk
10 oz. fish fry
Salt
Black pepper
Cayenne pepper
Oil for frying
Lemon juice

Enjoying gar surprised Rusty Munster.

Tip: In spite of its firm flesh when fresh, garfish does not freeze well. It becomes very soft.

Tip: Be sure to cut gar across the grain of the meat when cutting it into cooking-sized pieces.

Cut gar into vertical strips about 1½ inches wide. Whip the eggs and milk into an egg wash. In a separate bowl, mix the fish fry with salt, black pepper, and cayenne pepper to taste. Dip the fish strips in the egg wash, and then dredge them in the fish fry mixture. Fry the fish in vegetable oil at 350 to 375 degrees until golden brown. Squeeze lemon juice over the fish and serve. Serves 4.

Ruby's Hush Puppies

Hush puppies are a long-time Southern tradition. We have heard several historical theories on their origin, and they all involve feeding fried cornmeal to dogs to keep them quiet by New Orleans Ursuline nuns, an African cook in either Atlanta or New Orleans, confederate soldiers or Southern hunters and trappers. And to be honest, most recipes are pretty bland, perhaps fit only for the dogs. We got this one from Ruby Michel of Harvey in 1983. These are the best hush puppies that we have eaten. The fresh onions and garlic powder give them life, and the onions steam the inside of the hush puppies as they cook, keeping them moist. The sugar is Glenda's touch. The recipe is good with or without it.

2 cups cornmeal
1 cup flour
¼ cup sugar
¼ cup minced onion
½ tsp. garlic powder
1 egg
2 tbsp. canola oil
½ tsp. salt
2 tbsp. baking powder
1 cup milk
Oil for deep-frying

Tip: A small ice cream scoop works well instead of a spoon to dip batter from the bowl.

Tip: This recipe will not work if the baking powder is too old. To test your baking powder, add 1 teaspoon of it in ⅓ cup of hot water. If it foams, it has enough potency to do its job.

Mix all ingredients well with enough milk to make a heavy, soupy mixture. Use a tablespoon to scoop the mixture from the bowl and

another spoon to scrape it out of the first one and into the hot oil. Deep-fry at 350 degrees until golden brown. Be careful not to let the cooking oil get too hot. If it gets to 375 degrees, the outside of each hush puppy will become too dark, while the batter on the inside is still uncooked. Drain on paper towels. Serves 6

Bill Spahr's Fried Catfish Chips

Spahr's Seafood Restaurant uses a lot of hot sauce.

Des Allemands, Louisiana, named in French after the early German settlers of the area, was a small sleepy town of little note on the border of Lafourche and St. Charles parishes until catfish put it on the map. Two men, Father William McCallion, of St. Gertrude Catholic Church and founder of the Louisiana Catfish Festival, and Bill Spahr owner of Spahr's Seafood Restaurant, are recognized for establishing Des Allemands as the Catfish Capital of the World. Spahr's Seafood Restaurant is now famous even beyond Louisiana's borders for its fried catfish.

Bill offers some tips for frying perfect catfish every time. First, be sure to use a mixture of half cornmeal and half wheat flour to get the perfect coating. Second, hot sauce is necessary, both to firm the flesh of the fish and for flavoring. Not just any kind will work, though Cajun Chef is Bill's favorite brand, but he says that any thick hot sauce that doesn't tend to separate vinegar from pulp is good.

Third, he says that a person should know his fryer. If the oil doesn't sizzle immediately after the fish is added, the fryer has been overloaded and the fish will come out soggy. Less fish should be added at a time. Keep experimenting until you find how much fish the fryer can handle in a batch, then stick with the formula. Finally he directs not to over-fry the fish. "Most people," he says "fry their seafood to death."

1½ lb. catfish fillets
3 tbsp. hot sauce
1½ tsp. salt
½ tsp. black pepper
⅔ cup cornmeal
⅔ cup flour
Peanut oil for frying

Cut the catfish fillets diagonally into strips about the size of a man' finger. Place the catfish strips in a large pan. Add hot sauce, salt, and pepper to the catfish and toss to coat the catfish evenly. In a separat pan, mix the cornmeal and flour. Add a few pieces of catfish to the cornmeal and flour mixture and toss to coat. Repeat the process unti ll strips of catfish are coated. Heat oil in a deep fryer to 375 degrees. ry fish in small batches for approximately 1½-2 minutes. Don't verload the fryer. Cook until the fish are golden brown. Serves 4.

The Spahr's Seafood Story

Bill Spahr, the papa bear of Spahr's Seafood, was born in New Orleans on August 15, 1927. While Bill was in the U.S. Army, his parents moved to Des Allemands, where they had maintained a family camp. After discharge from the army in 1947, Bill joined his parents there and helped them run the Bayou Inn, the family business in nearby Raceland, Louisiana.

Bill and his wife, Thelma, went off on their own in 1968, building an Esso service station in a low marshy area between Des Allemands and Raceland. His vision was for a filling station, convenience store, and sandwich shop under one roof.

"They didn't have convenience stores in those days," he says. "Everyone thought I was nuts." But before long, trucks began using it as a pit stop because he fixed flat tires as well as providing all his other services. "Some mornings I would get up and see five or six eighteen-wheelers waiting for me with flats to fix and I knew I was going to have a busy day."

The cold cut sandwiches he sold at the business proved so popular that Thelma suggested selling hot sandwiches as well. Fried catfish was an easy choice for the sandwiches because catfish were common

Spahr's Seafood Restaurant has come a long way from two kitchen tables set up in an automobile service station.

in nearby Lac Des Allemands and because Bill fished for them commercially on the side.

So in 1969, he took his chicken fryer from his home to his business and what would become Spahr's Seafood Restaurant was born. After tiring of resisting repeated requests for fried catfish platters, the Spahrs relented and served them to customers on two kitchen tables set up in the service station.

The two tables grew to four, and then more. The word began to spread about "that service station on Highway 90 that serves the best fried catfish around."

Other seafood items were added. Thelma was from "down the bayou,"—Galliano, Louisiana—and the Spahrs could buy any seafood from there that wasn't available in Des Allemands. But fried catfish remained their signature seafood item.

"It was a big thing when we added French fries," remembers Bill. "A truck driver brought his family in on a weekend and one of his kids asked for French fries. So Thelma went to the store and bought some potatoes. We peeled them, cut them up, and fried them."

"We got known for our French fries," he goes on, "because we used fresh white potatoes. People would come from all around just to eat the French fries."

In the 1970s, seating maxed out at eighty chairs and people waited in line to get in. Bill removed the gasoline pumps from the front of the business, although their elevated concrete islands remained in place until 2006, when the new restaurant was built.

ot sauce is liberally doused over the catfish strips and ixed in by hand.

Hot sauce both seasons and firms up the catfish before frying.

In 2003, four years after Thelma died, the original Spahr's Restaurant burned to the ground and Bill retired. But his grandson, Donald Spahr, with two partners, opened a branch of Spahr's Seafood in Thibodaux in 2004, followed by another in the Ramada Inn in Houma in 2005.

In 2006, the partners rebuilt Spahr's Seafood to seat 140 at the original location on Highway 90, west of Des Allemands. With its rebirth, Bill "semi un-retired" and began coming in three or four days a week at 4:00 A.M. to cook the restaurant's famous seafood gumbo.

Bill attributes much of the phenomenal success of Spahr's Seafood to their close attention to the cooking process and to the quality of the seafood they use. "We watch our cooking real close. Most people who work in kitchens will overcook seafood.

"All we use is wild catfish, too," he added. "Never farm-raised. Local fish fillets are firmer and have a pinker color. Farmed catfish are white because they are washed so much. They even cook different. It's hard to describe the flavor of wild catfish in words; you just have to taste them."

Bill Spahr is still active and enjoys a good sense of humor.

Pescado Frito Con Tomate

This really good dish comes from Luis Campos, a native of Guatemala. When he told me how to cook it, he explained that it is a very common dish there—"Everybody cooks it." We tried it and we loved it. I asked him what the name of the dish was. "Fish," he replied. I kept nagging him for a name, so he called his mother. She told him it was "Fish," as well. So I gave it a name. This dish is incredibly good when fresh local tomatoes are on the market. They have the perfect blend of sugar and acid.

 lb. fish fillets
 medium tomatoes, finely chopped
 medium onions, minced
 jalapeño peppers, seeded and finely chopped
 cloves garlic, minced
 cup cooking oil
 salt and pepper to taste

Wash fillets and set aside. Sauté the tomatoes, onions, jalapeño peppers, and garlic in a small amount of cooking oil over a medium-high heat. Add salt and pepper. In a separate pan, fry the seasoned fish fillets in the rest of the oil to a light golden color. Do not overcook! Spread the fillets in a pan and cover with the tomato mixture. Simmer 10 minutes on top of the stove until the sauce blends with the fish. Serve plain or with cooked rice. Serves 4.

The Patriot

This recipe has a pedigree. Its creator, Larry Roussel, was honored by the Louisiana Senate in 2002 after he cooked it for "A Taste of the Senate." It took first place at the First Annual LSU Tailgate Cookoff in 2002, as well as first place in the seafood category at the North American Wild Game Cookoff in Lafayette. He has prepared it at numerous functions, always receiving accolades. Outdoors media personality Don Dubuc calls the recipe a "giant killer" because he has seen Larry knock off in competition some of the best chefs with their fancy portable kitchens. After Glenda watched Larry top off the red snapper fillets with jumbo lump crabmeat and large shrimp, she asked, "What's there not to like about this dish?"

Larry calls it the "Patriot" because of its color theme. The seven red snapper fillets represent his favorite president, Andrew Jackson, who was the seventh president of the United States. The fifty white shrimp represent the fifty states in the country and the pound of blue crab meat represents our one great nation. He cooks it somewhere on every patriotic holiday. The only hitch in this otherwise absolutely wonderful recipe is getting the LeGoût Cream Soup Base. It isn't available in stores in Louisiana. It can be ordered through the Internet or if one knows a restaurateur, he can easily get the product from his food service distributor. It's worth the effort.

Food has replaced fishing as the focal point in Larry Roussel's life. He cooks for play, he cooks for events, and he cooks for work.

The only challenging part of this recipe is getting the LeGoût Cream Soup Base.

2 sticks butter, divided
7 8-oz. red snapper fillets
Paul Prudhomme's Blackened Redfish Magic seasoning
1 cup water
1 1.75-oz. bag LaGoût Cream Soup Base
50 large peeled white shrimp
2 pinches salt, divided
1 lb. jumbo lump blue crab meat, drained

Red Snapper

Melt 1 stick of butter in a skillet. Dry the red snapper fillets with paper towels and thoroughly coat them with melted butter. Generously sprinkle the blackened redfish seasoning on both sides of each fillet. Place the fillets in the hot skillet and pan fry for 3 minutes, flip, and pan fry the other side for 2 minutes.

White Shrimp

Pour 1 cup water into another skillet. Using a whisk, gradually blend the cream soup base into the water until the mixture is smooth (about 1 minute). Add shrimp and 1 pinch of salt. Cook over medium heat for 10 minutes, stirring often.

Blue Crab

Melt 1 stick of butter in a skillet. Place crabmeat into the melted butter. Add 1 pinch of salt. Do not stir. Tilt pan at an angle with the crabmeat at the top and spoon the melted butter over the crabmeat just until crabmeat is warm. Drain and reserve the meat.

Assembling the Dish

Place the red snapper fillets onto red, white, and blue plates. Generously spoon the white shrimp sauce onto the fish fillets until each fillet is entirely covered. Top the entrée with jumbo lump crabmeat. Garnish with an American flag on a toothpick. Serves 6 plus the cook.

Crispy-Crunchy Oven-Fried Fish

Oven-fried fish are really misnamed as they are not fried at all, but instead are baked. They get their name from the delightfully crispy crust they develop if properly prepared, but they don't have the oi of fried fish. If you follow this recipe exactly, it is foolproof perfect Don't be scared of the horseradish in the dish; in this recipe, it doesn' have the pungent bite so often associated with it. One caution: It is important to use fillet portions that are 1 to 1½ inches thick. Thir fillets will dry out. The best fish species to use with this recipe are those with a flakey white flesh. A wire baking rack is necessary to ge all sides of the fish crispy. This will become a family favorite.

1½ lb. 1½-inch-thick fish fillets
4 slices white bread, torn into small pieces
2 tbsp. melted butter
Salt
Black pepper
2 tbsp. minced fresh parsley
1 green onion, minced
2 eggs
2 tsp. prepared horseradish
3 tbsp. mayonnaise
½ tsp. paprika
¼ tsp. red pepper
All-purpose flour

Wash the fillets and cut into 4-inch pieces. Set them aside in the refrigerator. Preheat oven to 350 degrees with the oven rack in the middle position. Coarsely chop bread in a food processor with abou 8 1-second pulses. Add melted butter, ¼ tsp. salt, and ¼ tsp. blac pepper and stir to coat the breadcrumbs. Spread the breadcrumb on a baking sheet and bake until deep golden brown and dry, abou 15 minutes, stirring twice while baking. After the breadcrumbs have cooled, toss them with the parsley and green onion. Increase the over temperature to 425 degrees. Whisk together the eggs, horseradish mayonnaise, paprika, red pepper, and ¼ tsp. black pepper. Whe smooth, whisk in 5 tbsp. flour. Spray a wire rack with non-stic cooking spray and set it on a rimmed baking sheet. Dry the fish pieces with paper towels and season them with salt and pepper. Dredge the fish in flour and shake off any excess. Dip each piece in the egg mixture, then coat all sides of the fish with the breadcrumbs. Use your hands to firmly press the breadcrumb layer to the fish, then place the finished pieces on the baking rack. Bake the fish 18-25 minutes until the crust is golden and the flesh flakes to the touch. Serves 4.

Nuts to You

This recipe has been in our files since early 1995, when I ran it in my LSU fisheries newsletter. We have no record of who submitted it. It has an absolutely wonderful nutty taste that accents the taste of the fish. We are very partial to crusting fish with nuts, and walnuts are a big favorite. Pecans would probably work just as well.

1½ lb. fish fillets
1 cup sour cream
1 tsp. celery salt
1 tsp. paprika
1 tsp. lemon juice
1 tsp. Worcestershire sauce
½ tsp. salt
¼ tsp. pepper
1 cup cracker crumbs
½ cup finely chopped walnuts

Rinse fillets and set aside. Combine sour cream, celery salt, paprika, lemon juice, Worcestershire sauce, salt, and pepper. Chill 30 minutes to blend flavors. In a separate bowl, combine cracker crumbs and walnuts. Dip the fillets in the sour cream mixture, then roll in crumb mixture. Place fish in a single layer on a well-greased 10x15-inch baking pan. Bake at 375 degrees for 18-20 minutes or until fish flakes easily when tested with a fork. Serves 4.

Crusty Onion-Baked Fillets

This mildly-seasoned dish brings out the best in the taste for any species of white-fleshed fish, saltwater or freshwater. Notice that there is no added salt and pepper. The dressing mix provides enough seasoning. As a bonus to its good taste, the dish is marvelously easy to prepare. The baked onion crust on the fillets makes an attractive presentation as well. We suspect that this recipe may work well with darker-fleshed fish such as Spanish mackerel and bluefish, but we have not tried it.

½ lb. fish fillets
½ cup sour cream
½ cup mayonnaise
½ package ranch-style dressing mix
2 3-oz. cans French fried onions

Wash the fillets and set aside. Combine sour cream, mayonnaise, and salad dressing mix in a medium-sized bowl. Crush the onions in a food processor or in a plastic bag with a rolling pin. Dip the fillets in the dressing mix then roll them in the crushed onions. Place the fillets in one layer in an ungreased baking dish in one layer. Bake in oven at 350 degrees for 20-25 minutes or until fish flakes easily with a fork. Serves 4.

Baked Bluefish

(Courtesy Bob Dennie)

This hard-fighting fish has a poor reputation among Louisiana cooks. Interestingly, it has a reputation as a good, if not prized, food fish on the Atlantic coast. Some describe the taste as strong or even "gamey." Anyone that has eaten smoked bluefish agrees that they are excellent this way, but the fish is more versatile than that. Properly handled, the gray flesh color, so off-putting to many people, cooks up near white. Some "experts" recommend gutting and bleeding the fish as soon as they are caught, a bloody and messy job most anglers don't want to do on their boats. Instead, proper prepping of the easy-to-fillet fish should do the trick. Simply discard the dark streak of flesh that runs through the middle of each fillet. If you cut all the way through the fillet on both sides of the streak, to remove the entire strip, the narrow strip of troublesome bones in this area will also be removed. Each fillet will produce two fillet strips.

2 lb. bluefish fillet strips
Salt and black pepper to taste
½ cup olive oil
½ cup Parmesan cheese
6-8 slices white bread
¼ cup chopped basil
¼ cup chopped parsley
1 tsp. salt
½ tsp. cayenne pepper
Lemon slices

Place the fillets in a baking dish. Salt and pepper them and baste with olive oil. Mix all of the other ingredients with the remainder of the olive oil and place on top of the fish. Bake 30-40 minutes at 350 degrees. Serve garnished with lemon slices. Serves 4.

Baked Mackerel with Mustard Butter

Both king and Spanish mackerel end up in Louisiana catches. Neither are white-fleshed mild fish and so end up being considered "difficult to cook." Mackerels are stronger tasting than many species of fish, especially king mackerel, which is also darker than its Spanish cousin. This recipe is designed for Spanish mackerel but will work with kings as well. The yellow mustard in the recipe has enough backbone to balance the assertive taste of the mackerel.

2 lb. mackerel fillets
3 cups water
¾ cup plus 3 tbsp. lemon juice, divided
6 tbsp. melted butter
4 tsp. yellow mustard
1 tsp. salt
½ tsp. black pepper
½ tsp. Old Bay Seasoning
½ tsp. paprika
Chopped parsley for garnish

Marinate fish in refrigerator for 20 minutes in water and ¾ cup lemon juice. Combine 3 tbsp. lemon juice, butter, mustard, salt, pepper, Old Bay Seasoning, and paprika and mix well. Place fish fillets in 10x15-inch baking dish. Brush generously with the mixture. Bake in a 350-degree oven for 4-6 minutes. Baste once more with mixture and bake 4-6 minutes more, or until fish flakes easily with a fork. Warm remaining mix, pour over fish, and sprinkle with chopped parsley before serving. Serves 4-6.

Spanish mackerel are a delicious and heart-healthy species.

Baked Fish on Board

We got this recipe from Allen Wiseman of Lafitte, Louisiana, in 1986. Now deceased, Allen was an ambitious commercial shrimper who owned, both single and in partnership, more offshore shrimp boats than anyone else between Lafitte-Barataria and Grand Isle. Probably most imaginatively named and symbolizing the one-time might of the shrimp fleet was his *Cajun Power*. The sheer size of the fillet needed for this dish limits it to species that grow large, such as red snapper, redfish, black drum, or grouper. In those pre-turtle-excluder-device, pre-finfish-excluder-device days, shrimpers could sort through their catch for a fish of suitable size to feed a hungry crew of five or six men. The dish looks "too tomatoey," but isn't. Nor is it too spicy.

Large redfish work well in the recipe.

- 3-5-lb. fish fillet
- large onion, cut in rings
- large bell pepper, cut in rings
- ½ cup vegetable oil
- salt and pepper, to taste
- lemon sliced
- ¼ cup sliced jalapeño peppers
- pt. spaghetti sauce

Rinse the fillet and set aside. Make a bed of onion and bell pepper rings in a heavy pot or roasting pan large enough for the fillet. Drizzle oil over the vegetables and lay the fillet on top. Salt and pepper to taste and place lemon slices over the fish. Bake covered at 400 degrees for 15 minutes. Remove from oven and scatter jalapeño peppers over the fillet. Cover with spaghetti sauce and bake an additional 30 minutes uncovered. Serves 8.

Fish and Banana Stew

This creation came from our family hunting and fishing camp in the Atchafalaya Basin. In fact, it, along with a couple of others, became the signature dishes of the camp. Glenda's brother, Dicki Ray, really loved it, as did our good friend and camp gourmet, gourmand, Sandy Corkern. We always made it with sac-a-lait, but it will work with almost any mild-tasting, white-fleshed fish from bass to sea trout to groupers. We even tried it with gaspergou and it was good. Bream fillets are a little small for this and will get lost in the sauce, even though it is mild. I would also avoid catfish because of their tendency to curl when cooked.

It doesn't look like a stew when it's done in a fancy style, but when we cooked it for a bunch of famished hunters or fishermen, we made it in huge pans and just covered the whole layer of fish with an even layer of the cream and banana sauce. The outdoorsmen were not dainty about serving themselves, lifting out one fillet at a time. Instead, they grabbed the biggest serving utensil in the camp and gouged up big chunks of it, which they unceremoniously dumped in their plates. It looked like the devil—a "mish-mashy" stew, but it tasted divine—and still does.

You like fish, right? You like bananas, right? What's there not to like? This is a genuine four-star recipe.

1½ lb. fish fillets
Salt and pepper to taste
1 lemon, juiced
6 tbsp. olive oil
1 cup chopped green onions
4 strips bacon, diced
6 tbsp. white wine
2 cups heavy cream
2 medium bananas, sliced
2 tsp. dried thyme
2 tbsp. white Worcestershire sauce

Season fillets with salt and pepper. Squeeze the juice of ½ lemon over the fish and sauté them in olive oil until they are flakey and done n a separate pan, sauté the green onions and bacon until golden. Deglaze pan with wine and cream. Reduce the cream on low heat until it is slightly thick and add the rest of the lemon juice, bananas, thyme, and Worcestershire. Spoon the sauce over the fillets. Serves 4.

Trout Amandine

This is our single-most favorite recipe for speckled or white trout. Glenda has cooked it for well over thirty years. When I bring a mess of trout home, we usually eat it fresh every other day for a week. This is always, without fail, the first one that we do—then we do a variety of recipes with the fillets that are left. This is simple! This is good!

3 lb. trout fillets
¾ cup milk
1 cup flour
Salt and pepper to taste
1½ sticks butter, divided
½ cup slivered almonds
2 tbsp. lemon juice

Dip fish in milk and dredge in flour that has been seasoned with salt and pepper. Melt half of the butter in a skillet and lightly brown the fish fillets over medium heat. Place the cooked fillets in an oven set on warm. Wipe the skillet and melt remaining butter. Lightly brown the almonds. Stir in the lemon juice and pour over the fish. Serves 6.

Trout with Shrimp Sauce

If you really want to impress someone, this is the dish to cook! It is one of the few dishes (along with amandine) that Glenda and I really prefer sea trout for. It works equally well with white trout, as speckled trout and any size fillets can be used. With small fillets, just put a small amount of the decadently delicious shrimp sauce on them. With thick fillets, just spoon on more. Since we conjured this up well over thirty years ago, it has held steady as one of our family favorites. It's even delightful as a leftover. We always keep a couple of cans of cream of shrimp soup in the cabinet so that we are ready to go.

1½ lb. trout fillets
Blackened redfish seasoning
Mayonnaise
Butter
1½ cups sliced fresh mushrooms
1 lb. peeled small shrimp tails
1 cup chopped green onions
1 cup minced parsley
2 10-oz. cans cream of shrimp soup

Rub the trout fillets on both sides with the blackened redfish seasoning and lay them on wax paper or a large pan. Spread mayonnaise on each fillet and set aside for 15 minutes. Place a non-stick frying pan on high heat. Add enough butter to barely cover the bottom of the pan when melted. When the butter is hot enough to sizzle when a drop of water is added, add the fillets in a single layer and cook until golden bronze in color. Add butter as needed until all of the fillets are cooked. Set the fillets aside. In the same pan, toast the mushrooms in butter until golden. Add shrimp, onions, and parsley and cook on low heat until the shrimp are pink and tender. Add cream of shrimp soup and blend well. Spread fish fillets on a baking pan. Spoon the shrimp mixture over the fillets and bake uncovered in a 375-degree oven for 15-20 minutes. Serves 4.

Mom's Zesty Creamed Fillets

Lane Lemaire is a dedicated speckled trout angler.

Tip: Lemon zest makes the dish, but don't overdo it at first. Test as you make the dish and add more if needed after tasting but before putting it in the oven.

Lane Lemaire of Lafayette credits his mom, Maggie Lemaire, with this recipe. "She is a simple Cajun cook from Gueydan," he says, "using salt, pepper, onions, bell pepper, celery, garlic, garlic powder and salt and pepper. None of that fancy stuff—and it is really good." Lane cooks like that, but he does a lot more. He loves to experiment. When he goes out of town on vacation, while the others shop for souvenirs, he patrols the local grocery stores looking for anything unique to try. "That's my gig," he chuckles. "I shop the grocery stores." Lane has been cooking since he was fifteen and learned through camp life. "Somebody was always cooking. My job was to chop seasonings while the men cooked and then do the dishes. But I watched. I love to eat. I like to see the expression on peoples' faces when they say, 'Man, this is good.'"

8 large trout fillets, or 16 small pieces
1 lb. peeled medium shrimp
Lemon pepper
Tony Chachere's Original Creole Seasoning
Salt to taste, or Mrs. Dash Original Blend
Olive oil
¾ cup chopped onions
¼ cup chopped celery
1 small red bell pepper, chopped
1 small yellow bell pepper, chopped
½ cup chopped green onions
½ cup chopped parsley
1 can cream of chicken soup
1 pt. half-and-half
2 lemons, zested and juiced
Garlic powder to taste

Spread fillets and shrimp separately on 2 pieces of foil. Sprinkle them with lemon pepper, Creole seasoning, and salt. Fold the fish and shrimp in the foil and set them aside to marinate while preparing the sauce. Heat a large pan with enough olive oil to coat the bottom. Add onions and celery and sauté until soft. Stir in bell peppers, green onions, and parsley. Add chicken soup and half-and-half and bring to a simmer. *Do not boil.* Add lemon juice, garlic powder, and zest of 1 lemon. Check seasoning and adjust as necessary. Spray a 10x13-inch pan with non-stick spray. Layer ½ of the sauce in the pan. Spread the fish and shrimp on the sauce. Pour the remainder of the sauce over the seafood and cover. Bake at 350 degrees for 45 minutes. Uncover the dish and bake an additional 15 minutes. Serves 8.

Pretty Snapper

This recipe comes to us from Woody Crews of Metairie. Some call him an insurance executive, but he is really a duck hunter, rig fisherman, and blue water marlin troller. Like the name says, the dish is pretty. The two pink-skinned red snapper fillets creating a nest for the mound of crabmeat dressing looks irresistible. And it tastes better than it looks—luscious flakes of red snapper mixed with juicy crabmeat on a fork is pure delight. "I like to cook this when I have clients over as guests," explains Woody. The dish was inspired by his mother, Joyce Lafaye Crews, who used to bake whole croakers in the 1970s when so many large ones were available. Woody cooks this dish with fillets instead of whole fish because he wants it to be kid friendly—without bones. Woody learned to cook at hunting camps by watching other cooks. "Presentation wasn't nearly as important to a bunch of hungry hunters," he explains, "as were flavor and volume."

2 2-lb. skin-on, scaled red snapper fillets
2 tbsp. olive oil
4 tbsp. butter, divided
¼ bell pepper, finely chopped
1 stalk celery, finely chopped
2 shallots, finely chopped
3 green onion bottoms, chopped
1 lb. lump crabmeat
Kosher salt to taste
Black pepper to taste
1 cup breadcrumbs
3 tbsp. heavy whipping cream
Crisco shortening
2 lemons, divided
1 tbsp. parsley, chopped

Remove any pin bones along the lateral line of each fillet with a pair of pliers. Rinse the fillets and pat them dry with paper towels. Put 1 tbsp. olive oil in a medium skillet, add 1 tbsp. butter, and heat until the butter has melted. Add bell pepper, celery, shallots, and green onion bottoms. Sauté until tender. Add crabmeat, salt, pepper,

The amiable giant Woody Crews learned his cooking skills at hunting camps.

Tip: The shallots called for in this recipe are not green onions but rather true shallots, which Woody prefers for their "less oniony" taste. He says he does not use the green onion tops because he finds them slimy.

breadcrumbs, and whipping cream. Mix well. Taste to be sure salt and pepper are okay. Salt and pepper the fillets to taste. Grease a large baking dish with shortening. Mound the crab mixture in the middle of the dish. Rub olive oil over the skin of the fillets. Place the fillets around the crab mixture skin side up. Drizzle the juice of 1 lemon over the fillets and crabmeat. Slice the remaining lemon and scatter the slices on the fillets. Cut remaining butter into smaller pats and place on top of the crabmeat mixture. Bake at 350 degrees for 15 minutes. Baste the fillets with butter from the bottom of the pan 2-3 times during baking. Turn oven to broil and broil until the fillets begin to brown. Sprinkle with parsley before serving. Serves 8.

Kane's Honey Parmesan Fish

As usual, Mary Poe generously gives credit to someone else for her cooking.

Tip: To add a smoky taste, Mary often gets Jeff to grill the fish for the first 10 minutes of cooking time before finishing it in the oven.

Tip: Three- to four-pound fish produce the right-sized fillets for generous individual servings for this recipe.

Mary Poe and her husband, Jeff, own Big Lake Guide Service on the eastern shore of Calcasieu Lake in southwestern Louisiana. Mary actively guides anglers in pursuit of speckled trout and redfish. But she is also known for her cooking, much of the credit for which she passes on to her late father, Major Newlin, who apparently loved to cook as much as he loved to fish. Mary says that she got this recipe from Kane Mitchell, a friend from Sulphur. She chatted him up at the bank where he works. Mary calls Kane's dish "absolutely the best" and cooks it often.

4 redfish or other firm-fleshed, skin-on, scale-on fillets
⅔ cup honey
Tony Chachere's Creole Seasoning to taste
⅔ cup sour cream
⅔ cup mayonnaise
1 tbsp. sherry
1½ cup grated Parmesan cheese
1 tbsp. lemon juice

If the rib cages have not been removed from the fillets, cut them out, and discard them. Wash the fillets and pat them dry. Spread honey over the flesh side of the fillets. Season them liberally with Creole seasoning. In a small bowl, combine sour cream, mayonnaise, sherry, Parmesan cheese, and lemon juice. Spread the mixture thickly over the fish fillets. The mixture needs to stand up on the fillets. If it wants to run off the first one, add more cheese. Bake for 30 minutes at 350 degrees or until browned. Serves 4.

Heavenly Fillets

This one is appropriately named and has been in our collection for seventeen years. Many versions of it exist, some of them older than ours. We have used it often with offshore fish, inshore saltwater fish, and freshwater fish. It is good with them all. We suggest, however, that it not be used on very small fillets, as it is too difficult to sauce them properly. The Parmesan cheese is wonderful in this preparation, but not overwhelming. Using Louisiana-style hot sauce instead of granulated pepper is perfect.

2 lb. fish fillets
2 tbsp. lemon juice
½ cup grated Parmesan cheese
¼ cup butter
3 tbsp. mayonnaise
3 tbsp. chopped green onions
½ tsp. salt
4 dashes hot sauce

Place fillets in a single layer in a well-greased baking dish. Brush with lemon juice. Broil the fish 4 inches from the heat source for 5 minutes or until the fish flakes easily when tested with a fork. Remove from heat. While the fish is broiling, combine the remaining ingredients and spread it over the fish. Broil 2 more minutes or until lightly browned. Serves 4.

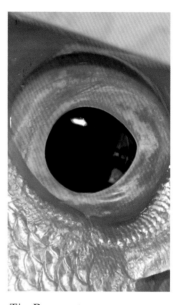

Tip: Be sure to scoop up any sauce that runs off the fish and serve it with the fish at the table. It is far too good to waste.

Marlin Delight

(Courtesy Myron Fischer)

Louisiana has two species of marlin, the blue marlin and the white marlin, both of which make quite good table fare. Classified as game fish in the Atlantic Ocean, which includes the Gulf of Mexico, these billfish cannot be purchased here. A modest number of blue water sportfishing specialists do troll for them off our coast and catches can be quite good. Nowadays most of these fish are released and not brought to dock to be photographed, as they once were. Occasionally one of these hard fighters will actually fight itself to death or a shark will attack and kill it while it is hooked. The flesh of these animals is just too good to be wasted. This recipe comes from Joe Yurt, who was the official club biologist and weighmaster for the New Orleans Big Game Fishing Club, headquartered at Port Eads on the South Pass of the Mississippi River. It was printed in the club newsletter, *Mouth of the River*, in 1975. These fish compare favorably to swordfish on the table; don't ruin them by overcooking them.

4 6-8-oz. marlin steaks or fillet pieces, ¾ inch thick
1 cup milk
2 tbsp. butter, melted
3 tbsp. lemon juice
2 cloves garlic, minced
2 tsp. chopped parsley
Salt and pepper, to taste
Panko breadcrumbs

Remove any dark flesh from the fish and discard. Marinate the fillets in milk in the refrigerator for 1 hour. Mix butter, lemon juice, garlic and parsley. Remove the fish from milk, dip in the butter mixture and roll in the breadcrumbs. Bake at 350 degrees for 15 minutes. Test the fish for doneness by flaking a fillet with a fork. If the fish flakes easily, it is done. Do not overcook! Serves 4.

Citus Therapy

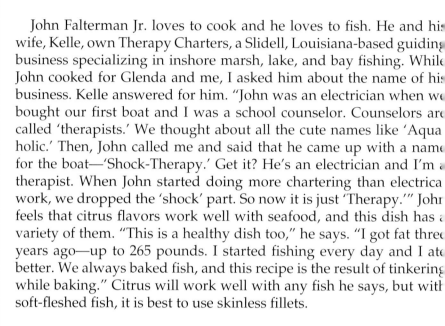

John Falterman Jr. loves to cook and he loves to fish. He and his wife, Kelle, own Therapy Charters, a Slidell, Louisiana-based guiding business specializing in inshore marsh, lake, and bay fishing. While John cooked for Glenda and me, I asked him about the name of his business. Kelle answered for him. "John was an electrician when we bought our first boat and I was a school counselor. Counselors are called 'therapists.' We thought about all the cute names like 'Aqua-holic.' Then, John called me and said that he came up with a name for the boat—'Shock-Therapy.' Get it? He's an electrician and I'm a therapist. When John started doing more chartering than electrical work, we dropped the 'shock' part. So now it is just 'Therapy.'" John feels that citrus flavors work well with seafood, and this dish has a variety of them. "This is a healthy dish too," he says. "I got fat three years ago—up to 265 pounds. I started fishing every day and I ate better. We always baked fish, and this recipe is the result of tinkering while baking." Citrus will work well with any fish he says, but with soft-fleshed fish, it is best to use skinless fillets.

John Falterman says that his love of cooking started in the Boy Scouts because they were always creatively cooking something over a fire.

Tip: You may also cook this dish on an outdoor barbeque grill instead of in the oven. Grill for 25-30 minutes.

2 15-18-inch skin- and scales-on redfish fillets
2 oranges, divided
2 lemons, divided
2 limes, divided
1½ tsp. Creole seasoning
1½ tsp. lemon pepper
½ tsp. dried oregano
½ tsp. dried thyme
1 stick butter, softened

Place the fillets scale-side-down on a baking pan. Zest 1 orange, 1 lemon, and 1 lime over the fillets. Sprinkle Creole seasoning, lemon pepper, oregano, and thyme over the fillets. Slice 1 orange, 1 lemon, and 1 lime and alternate slices on top of the fish. Scatter ¼-inch slices of butter over the top of the fish. Bake at 400 degrees until lightly browned and the meat begins to flake. Do not let the fish dry out. Add butter as needed to the top of the fish. Serves 4.

Redfish T-Rue

Tina Rue has a cabinet full of spices and knows how to use every one of them.

Tip: This dish can be made with any kind of fish. If she uses skinless fillets, Tina will dip the fillets in butter rather than brush them, and then she will roll them in the seasonings.

Tip: Dried herbs are used because the oven's heat tends to singe larger pieces of fresh herbs.

Tip: Tina uses chopped pecans more often than almonds. Either works well.

Tina Rue (T-Rue or Rue-Rue to her friends) routinely cooks with her assistant and lodge manager, Chrystal Fontenot, at the Calcasieu Charter Service Lodge on the shore of Lake Calcasieu, which she and her charter captain husband, Erik, own. She cooks twice a day during the seventy day waterfowl hunting season and three to four days a week the rest of the year except February—their vacation month. What do they do in February? Go somewhere else and fish, that's what! When they got married at Isla Mujeres, Mexico, thirteen years ago, it was the day after catching twenty-one sailfish. They got married early in the morning, so that they could fish that day too. Tina has been using this dish for eighteen years, and it has become a staple item at the lodge. She doesn't like to fry. For her customers who usually only like their fish fried, this has a little crunch that they like.

4 12-14-inch skin- and scale-on redfish fillets
1 stick butter
1 tbsp. lemon juice
1 cup plain breadcrumbs
½ tsp. mustard powder
½ tsp. garlic granules
½ tsp. Parmesan cheese
½ tsp. dried dill
½ tsp. dried rosemary
Creole seasoning to taste
1 lemon, sliced
½ cup slivered almonds
2 tbsp. red peppercorns

Wash the fillets and pat them dry with paper towels. Place dried fillets in a baking pan that has been sprayed with non-stick cooking spray. Melt the butter and add lemon juice. Brush the fish fillets with the lemon-butter mixture. In a separate bowl, combine breadcrumbs, mustard powder, garlic, Parmesan cheese, dill, rosemary, and Creole seasoning and sprinkle over the fish fillets. Place 1 slice of lemon on top of each fillet and sprinkle with almonds and red peppercorns. Bake in oven at 350 degrees for 25-30 minutes. Serves 4.

Basil and Lemon-Butter Broiled Flounder

Flounders are amongst the leanest and mildest-flavored fish in Louisiana. This makes them one of the most popular saltwater fish in the state. They also tend to be a little dry-fleshed, something that can easily be overcome with proper cooking. Flounder begs to be broiled with butter and the simplest of herbs and spices, and indeed broiled flounder is a classic preparation wherever flounder are cooked. Basil is the friend of fish and one of our favorite fish cookery herbs. Strangely, while we grow basil in large quantities and have it available fresh during all of the warm months, we prefer dried basil for this dish.

4 1-1½-pound flounders, headed, gutted, and scaled
2 sticks butter
Juice of 2 lemons
1 tsp. dried basil
2 tsp. salt
1 tsp. black pepper

Set the oven on broil. Wash the fish thoroughly and trim the tail fin. With a sharp knife, score the fish across the body to the backbone, making cuts 1 inch apart. Melt the butter in a saucepan and add the lemon juice, basil, salt, and pepper. When the oven reaches broiling temperature, spoon ½ of the lemon-butter into the cuts and place the fish in the oven. After 3 minutes, spoon the rest of the lemon-butter mixture over the fish. Broil an additional 3 minutes. Serves 4.

Tip: Do not reduce the salt called for in the lemon-butter. Much of it and the butter runs off of the fish during cooking rather than penetrating the flesh.

Orange-Broiled Fish

For years, Glenda and I would prepare this dish two to three times
month, all spring and summer, while I was chasing speckled trout.
Obviously it was a favorite, and it is simple, too. It makes use of the
wonderful fruity taste of oranges, but the oranges don't overpower the
ish. The only challenging thing for those who must have crusty fish
that pouring the remaining butter-orange mixture over the breaded
ish produces a softer crust. Glenda prefers that to withholding it be-
ause she says that not using it reduces the orange flavor too much.

½ lb. fish fillets
¼ cup butter, melted
tbsp. concentrated orange juice
tbsp. grated orange zest
cup finely crushed Ritz crackers
tsp. paprika
½ tsp. salt
¼ tsp. pepper

Cut the fish into serving-size portions. Combine butter, orange
juice, and zest. In another bowl, mix cracker crumbs, paprika, salt,
and pepper. Dip the fish in butter mixture and then roll them in the
cracker mixture. Arrange the fish on a broiling pan. Baste the fish
with any remaining margarine mixture. Broil 6-8 minutes or until the
ish is white and flakey. Serves 4.

Nancy's Fast Fish Delite

(Courtesy Brent Fay)

In the mid-1980s, I did a lot of exploratory fisheries work to develop new fisheries in Louisiana. One of the guys who I worked with a lot was a Florida transplant and an ace king mackerel and snapper fisherman named Bayward Stone. Bayward was one of the most knowledgeable people I have ever met and seemed to know intuitively what was going on in any fish's mind. I shared many meal with him and his girlfriend, Nancy, in one of the Estay Seafood Company camps on Grand Isle. This recipe is one of Nancy's. It is so simple and so good—almost too good for being so simple. Strange to say, I never knew Nancy's last name. Bayward is gone now, a victim of a heart attack, and I have lost track of Nancy completely, but her recipe lives on.

1 6-8 oz. fish fillet per person, ½ to ¾ inch thick
Oil
Seasoned salt to taste
Fresh vegetable medley (broccoli, cauliflower, carrots, onions, celery,
   mushrooms, bell pepper)
Black pepper to taste
Garlic salt to taste
Butter

Rinse the fish and set aside. Cut sections of freezer paper large enough to double over each serving of fish and form a package. Lightly oil the center of the unwaxed- side of the freezer paper, about the size of a fish fillet. Place each fillet on a piece of freezer paper and sprinkle with seasoned salt. Cut the vegetables into small pieces and place the vegetable medley on top of the fish. Season with seasoned salt, black pepper, and garlic salt. Place a few pats of butter on top of the vegetables. Fold the freezer paper over the fillet "package style." Cook each fillet separately in a microwave on the "high" setting for 10 minutes. If several servings are being cooked, keep the early ones in a warm oven until all are done or reheat in the microwave.

Smothered Gaspergou

Freshwater gaspergou closely resemble their saltwater cousin, the black drum.

Billie Jeansonne of Boyce, Louisiana, shared this recipe with us. first learned how good of a cook she is when I stayed with her and her husband, Bob, at their fishing camp on Cross Bayou. Bob sallies from the camp into the Saline-Larto lake and bayou complex to chase his favorite fish—white perch (sac-a-lait). Bob says that his father, who was "pure French," couldn't speak English when he started school "We ate smothered gou a lot growing up," he says, "probably once or twice a month on Fridays. On Thursday, he would go to the fish market and buy a twelve- or fifteen-pound gou. It took a big one to feed us six boys and Mom and Dad." Billie says that because Bob enjoys stewed gou so much, she borrowed this recipe from the "little bitty" cookbook, *Historic and Authentic French Recipes of the Marksville Area.* Billie is modest for being such a good cook. "I was raised in the country and I'm still a country girl and I'm seventy now. Mother cooked west Texas style—white gravy and biscuits. She never did season a lot, salt, pepper, and a few onions. We didn't know what these other seasonings were for." She learned!

1 4-5-lb. gaspergou, scaled and cleaned
Salt and pepper to taste
2 tbsp. oil
1 large onion, chopped
1 medium bell pepper, chopped
1 clove garlic, minced
1 bunch shallots, chopped
½ cup chopped parsley

Score the fish with a knife at its thickest points. Liberally season with salt and pepper and set aside. Pour oil in a heavy cast iron Dutch oven. Add onion, bell pepper, and garlic and sauté until limp. Add shallots and parsley. Lay the fish on top of the seasonings. Cover the pot and simmer for about 45 minutes. Turn the fish midway through cooking and place some of the seasonings on top. The fish should flake easily but not fall apart when cooked. Serve with rice for the gravy. Serves 4.

Gar Roast

Gayle Daigle, who cooked this recipe for us, says that when she was growing up in Duson, Louisiana, her mother would cook this. "It was a once-in-a-while thing, not something we ate as common a catfish. But during Lent we ate more fish. We needed different ways and kinds of fish to prepare. This was a treat!" It is indeed a real treat and it is "real Cajun," from the simplicity of ingredients down to its great brown gravy to serve over rice. The texture of the garfish in a roast is unique—very, very similar to pork loin. You wouldn' know that you are eating fish if you were blindfolded. Her husband Tommy's family all cooked garfish too when he was growing up. "Everybody cooks. I'm a Cajun girl," she says. "Our lives are focused around food when we get together. It's our culture where we lived and where we live now. We like to catch our own! Fish! Hunt!"

1 4-lb. gar roast
1 medium onion, chopped, divided
9 cloves garlic, minced, divided
3 tsp. chopped fresh parsley, divided
1 green onion, chopped
Salt
Red pepper
4 tbsp. olive oil
2 tbsp. flour
½ cup water

Insert the knife deeply into one side of an end of the loin to make a pocket. Repeat the process for the other side of the same end. Then flip the roast over and repeat the process to make two pockets on that end. Set the loin aside. In a medium bowl, combine half the onion, 8 cloves of garlic, 2 tsp. parsley, green onion, 1 tsp. salt, and ½ tsp. red pepper. Mix well and stuff each pocket of the roast. Season all sides of the roast with salt and red pepper. Pour the olive oil in the bottom of a cast iron Dutch oven or heavy stainless steel pot. Add the roast and brown it on all sides over medium-high heat. Add water as needed to prevent the roast from sticking. Reduce heat to low and cook for 1 hour. Remove roast from pan and set aside. Add ½ onion, 1 clove

Gayle Daigle cooks her roast on top of an outdoor burner in a black iron pot.

arlic, 1 tsp. parsley, and flour. Sauté until vegetables are tender and
he flour is brown. Add water and cook until gravy has thickened.
adle gravy around roast and serve with rice. Serves 8.

Gar Balls

Mike DeSoto calls garfish part of his Avoyelles Parish Cajun heritage.

Tip: If you buy un-ground garfish meat, Mike advises to probe the meat with a finger. If the finger goes into the meat, the gar drowned in the net and the meat will be mushy. "Anyone who sells garfish knows what you are doing and expects it," he says.

Tip: Do not overmix the gar balls because they will become too dense.

Rusty Munster of Slidell, Louisiana, imported help for this one. His culinary heavy cannon is his long-time friend Mike DeSoto of Hessmer, a small town near Marksville, Louisiana. A couple of nights before, Rusty and his pal Ray Ohler, with assists from Todd Melerin and Pete Covignac, all of Slidell, made a bowfishing trip in the canals of the upscale waterfront developments of Oak Harbor and Eden Isles. Mike is here to help Rusty cook up the prehistoric fish they arrowed. Rusty cooks and Mike consults and narrates.

Avoyelles Parish, Mike's home parish, is the northernmost Cajun parish in Louisiana and the people there cling strongly to their traditions. And one of their traditions, he says, is eating garfish. "On Friday night that's all you eat—garfish. Most people elsewhere call it a trash fish. I went to a fishing rodeo in Texas and they threw them in the dumpster after the rodeo. In Avoyelles Parish, they would have shot you for that. In the *Journal,* the newspaper in Avoyelles, they carry photos of fathers and kids with garfish like other places post deer. That's every week!

"My parents grew up in the community of Belledeau on Choctaw Bayou. My grandfather, Joe DeSoto, told me how important garfish figured in our family's lives. The men would corral them in the shallows of the bayou. When one tried to pass, they would hit it on the head with an axe." They were doing this to feed their families. This wasn't sport fishing.

"A lot of people in Avoyelles Parish eat a lot of garfish. They sell it out of the back of pickup trucks at the farmers' market across from the school board office. When I was in elementary and high school, they served garfish balls and fried garfish every Friday. You could pick either one," he says with a mischievous grin. "If you were nice, you could get both."

Gar balls, Mike says, are always eaten with gravy and rice. "Back in the day, when people didn't have enough garfish, they cut the fish with mashed potatoes but never more than 20 to 25 percent potatoes. Back then, people would scrape the flesh from its connective tissue by hand with a spoon. But nowadays, most people use ground garfish

lb. ground garfish
cup diced green onions
cup diced parsley
¾ cup chopped onions, divided
reole seasoning
¾ cups flour, divided
tsp. salt
tsp. pepper
il for frying
clove garlic
qt. water
tsp. crab boil

Alligator gar come equipped with impressive dental equipment and an armored body.

In a large bowl, use your hands to mix the ground garfish, green nions, parsley, ¾ cup onions, and Creole seasoning to taste. Make alls about 2½-3 inches in diameter. Slightly flatten the balls. Mix 2 ups flour, salt and pepper. Dredge the gar balls in the flour mixture. o not shake off excess flour. Deep-fry the gar balls until golden rown. Discard all but ¾ cup of oil. Add 2 cups onions and garlic and auté until the onions have softened. Add ¾ cup flour and make a roux ver medium heat, stirring constantly until the flour has darkened to he desired brown color. Add water and stir to blend. Add crab boil nd gar balls and simmer 20-30 minutes to make the gravy. Taste for alt and pepper. Serve over rice. Serves 8-10.

Mae-Mae's Fish Sauce Piquant

Outdoor kitchens are a tradition in Cajun Country. "Most ever Cajun man wants one," says John Dupuis. "It's where they cook thei meals. Their wife cooks inside the house. That way they don't ge fussed at for making a mess. Just about every one of my hunting an fishing friends has one. My wife Dana hates helping me with thi dish. I tell her that she has to stir it every five minutes for four hour: Tomato sauce and paste have a very strong acid taste, so the longe you cook it the better it is. I've done it—cooking for only three hour: but it's not the same. You can taste it."

John defines a sauce piquant as "a thick tomato gravy," not muc different from a spaghetti sauce except for the addition of roux. A courtbouillon, he says is more saucy—more of a soup—a lot les tomato sauce and more liquid. A courtbouillon will have a lot more c the featured seafood's flavor in it and will taste more of fish than thi: "This gravy is more about tomato." The piquant taste comes from the Rotel tomatoes and the Creole seasoning. If you want peppery he says, you can use the hot version of canned Rotel tomatoes. Th basic sauce can be used with shrimp, turtle, or any kind of fish. John' personal favorite is catfish.

2 lb. fish fillets, cut into 3x2-inch pieces
Chop's Original Blend Cajun Seasoning, to taste
1 bell pepper, chopped
3 large onions, chopped
⅓ cup olive oil
1 15-oz. can diced tomatoes
1 10-oz. can original Rotel tomatoes
1 15-oz. can tomato sauce
1 6-oz. can tomato paste
2 tbsp. dry roux
1½ cups water

Cut fish into 2x3-inch pieces, sprinkle liberally with Chop's Caju seasoning and marinate overnight in a covered dish. Sauté the bel pepper and the onions in a Dutch oven in olive oil until tender and light brown. Add tomatoes, tomato sauce, and tomato paste to onior

John Dupuis says it is important to shake the pot rather than stir to avoid breaking up the delicate fish.

Tip: Dry roux is sold as a powder. At first glance, it appears too light in color to produce a dark brown color like traditional roux made with oil. John's personal preference is Kary's Roux, made in Ville Platte.

Tip: John loves Chops Cajun Seasoning, but if it is unavailable, use another Creole seasoning (sorry, John).

Tip: While John loves to cook in cast iron, he prefers to cook this dish in an aluminum pan to avoid the sauce picking up an "iron taste."

mixture. Stir in Chop's Cajun Seasoning to taste and 1½ cups water, cover, and simmer over medium heat for 1 hour, stirring often. Add dry roux and stir until well mixed. Simmer an additional 3 hours. Stir very often. If during cooking, the sauce starts to "pop" when it bubbles, the sauce is getting too thick, and water should be added 1½ cups at a time. After the 3-hour simmer, place each piece of fish in the sauce individually, one layer deep. Make sure each piece of fish is covered with sauce. Do not overfill. Cover the pot and cook ½ hour. Do not stir. Shake the pot every 10 minutes to prevent the fish from sticking. Serve with cooked rice. Serves 6-8.

Swas-a-la

Barbara Picard says that one of her earliest memories is of her grandmother Guillot cooking turtle sauce piquant with turtle eggs.

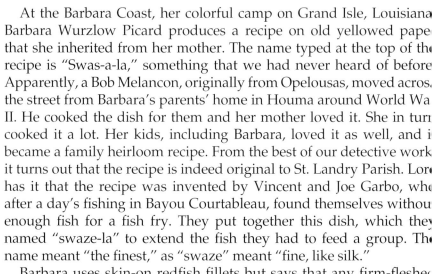

At the Barbara Coast, her colorful camp on Grand Isle, Louisiana, Barbara Wurzlow Picard produces a recipe on old yellowed paper that she inherited from her mother. The name typed at the top of the recipe is "Swas-a-la," something that we had never heard of before. Apparently, a Bob Melancon, originally from Opelousas, moved across the street from Barbara's parents' home in Houma around World War II. He cooked the dish for them and her mother loved it. She in turn cooked it a lot. Her kids, including Barbara, loved it as well, and it became a family heirloom recipe. From the best of our detective work, it turns out that the recipe is indeed original to St. Landry Parish. Lore has it that the recipe was invented by Vincent and Joe Garbo, who, after a day's fishing in Bayou Courtableau, found themselves without enough fish for a fish fry. They put together this dish, which they named "swaze-la" to extend the fish they had to feed a group. The name meant "the finest," as "swaze" meant "fine, like silk."

Barbara uses skin-on redfish fillets but says that any firm-fleshed fish, like cobia or amberjack, would be good. Les, Barbara's husband, suggests that it was likely that gaspergou would have been the fish of choice in St. Landry, an inland parish. Barbara thinks that the addition of olives was her mother's touch, but she calls them a necessary ingredient. The large amount of basil, the self-anointed "Basil Queen" admits, is her thing. In spite of the dish being red, she says, "It is not too tomatoey. It's like a vegetable garden."

5-6 lb. fish fillets
Salt
Red pepper
Black pepper
2 onions
5 cloves garlic
1 bunch green onions
1 bell pepper
1 bunch parsley
½ bunch celery, minced
1 sprig fresh basil
1 6-oz. can tomato paste
1 15-oz. can tomato sauce
1 cup fish stock or chicken broth
1 lb. fresh mushrooms, sliced
2 tbsp. olive oil
2 bay leaves
1 5-oz. jar olives
Creole seasoning

Wash fillets and season well with salt, red pepper, and black pepper. Set aside. Clean onions, garlic, green onions, bell pepper, and parsley and chop in a food processor. Add the celery to the onion mixture and set aside. Remove leaves from basil sprig and set aside. Mix tomato paste, tomato sauce, and fish stock. Put olive oil and bay leaves in a roaster and make a layer of fish, skin side down. Sprinkle a layer of the onion mixture, tomato mixture, olives, basil leaves, mushrooms, and Creole seasoning over the fish. Repeat until all fish has been added. Cover the roaster and put it over 2 burners on the stove with high heat until it begins to boil. Reduce heat to low and simmer about 1 hour until the fish flakes easily with a fork. Periodically shake the roaster to prevent sticking. *Do not stir!* Serve over rice or French bread. Serves 8-10.

Swas-a-la is a layered dish.

Smoked Mullet

John Supan uses a water smoker, minus the water pan, for this job.

Smoked fish are generally unappreciated in Louisiana. Striped mullet, so loved in the other Gulf states, are also not often an ingredient in Cajun and Creole cooking. So it's only natural that someone with roots elsewhere would be needed to introduce Louisianians to the gastronomic delight of smoked mullet. LSU fisheries professor John Supan, a native of West Virginia, lived and worked on the Mississippi coast a number of years, where he learned the secrets of this dish, sometimes called "Biloxi Bacon." He says he has smoked many kinds of fish, but that mullet is one of the very best. I agree! It has a mild, but rich taste and a firm, but slightly flakey, texture. Best results will be obtained by soaking the hickory chips overnight, so plan ahead. And, most importantly, do not allow the fillets to soak in the brine over the specified time.

4 cups hickory chips
2 lb. fresh mullet fillets
1 gal. water
1 cup salt
3 lb. charcoal briquettes
1 stick margarine, melted
1 tbsp. Worcestershire sauce

Soak the hickory chips in water at least 4 hours, preferably 8 hours before using them. Wash the mullet fillets. Make a brine with the water and salt. Add the fillets and set the pan in the refrigerator for 1½ hours. Do not allow them to brine longer or they will quickly become too salty. Light the charcoal briquettes in the smoker. Drain and rinse the fillets and mix them with the melted margarine and Worcestershire sauce. Set aside. When the briquettes are mostly white, spread the mullet fillets, red side down, on the grates of the smoker. Since the degree of doneness will be determined by color, having the white side of the fillets up is helpful. Add 2 or 3 handfuls of soaked hickory chips to the briquettes and cover the smoker. Whenever the smoke decreases, add more chips. Smoke until the fillets are completely golden in color but not brown, 40-45 minutes. Remove from smoker and serve. Serves 4.

Brother's Smoked Pompano

Brother Stipelcovich still has seawater in his veins at age eighty-four.

Tip: This recipe is also good for oysters, shrimp, or speckled trout.

Tip: If your smoker does not heat to 300 degrees, you will need to adjust your cooking time.

Lawrence "Brother" Stipelcovich, the originator of this recipe, is a legend in the commercial fishery of the Mississippi River Delta. Well over eighty years old, Brother is still spry and active, and spends as much time on boats as possible. When he cooks this, everybody comes over—family, neighbors, and friends. It's comical, watching them all standing with a plate in one hand and a gleam in their eyes. Pompano are the most expensive fish in the market today and have been for many years. This recipe uses an electric water smoker.

12 medium pompano skin-on fillets, scaled
2 sticks butter, divided
1 16-oz. bottle Italian dressing, divided
1¼ cup Dale's Steak Seasoning, divided
Salt, to taste
1 tbsp. minced garlic
3 lemons, divided

Remove the pin bones from along the center lateral line of each fillet. Melt 1 stick of butter in a medium saucepan. Add 1 cup Italian dressing and 1 cup steak seasoning and marinate the fillets in the mixture overnight. Layer the marinated fillets in a 10x14-inch pan. Lightly salt the fillets. Melt 1 stick butter in a medium sauce pan, 1 cup Italian dressing, ¼ cup steak seasoning, and minced garlic. Squeeze the juice of 2 lemons into the sauce mixture and thinly slice 1 lemon and add to the mixture. Mix well and heat until hot. Spoon the sauce mixture liberally over the fillets, arrange lemon slices over fillets, and cover with foil, folded loosely over the pan. Fill the water pan in the water smoker with water. Add 5 hickory chips that have been soaked overnight to the smoker and set the temperature to 300 degrees. After 45 minutes, baste the fish again with the juice from the pan. Fold the foil back in place and return the fish to the smoker to brown. Smoke the fillets for a total of 1 hour and 20 minutes. Serves 6.

The Making of a Fisherman

In February 1928, Lawrence Stipelcovich was born in Empire, Louisiana, to Joseph Peter Stipelcovich and the former Mary Louise Bowers. Joseph's father had emigrated from Croatia to Plaquemines Parish to become an oysterman, but he died shortly after arrival. The eight children of the family were reared by an older son.

Joseph was a shrimper most of his life, pulling a single-rig shrimp trawl from a gasoline-powered "putt-putt." Two of his brothers also became shrimpers and one became a shrimp buyer.

After bearing five children, the eldest and youngest of which died in infancy, Mary passed away when Lawrence, nicknamed "Brother," was seven years old. All three of the surviving children were boys and all three became shrimpers. "It was pretty hard without a momma," says Brother. "With Dad shrimping all the time, we pretty much raised ourselves."

The road to lower Plaquemines Parish was gravel and dirt when Brother was young, but he proudly recalls that his father was one of the first in that part of the parish to get a car. "Most people took a bus when they wanted to go to the city [New Orleans]" says Brother. "When the bus would get stuck in the mud around Lake Hermitage, the shrimp trucks hauling shrimp up the road pulled them out.

"I hunted and fished as a kid. Dad taught us early to hunt ducks and also taught us wood carving. He market-hunted ducks. They were packed in oil in barrels and went by train to the city. Crabs would go to city by train, too. No ice. We packed the live crabs with wet Spanish moss. We used cow ears to catch the crabs on trotlines.

"What ice we did get would come down by train. The blocks had to be hand-shaved."

Brother talks matter-of-factly of life in his youth, without waxing nostalgic. "All the people then had big gardens. A few had peach trees, and there was citrus in Boothville. If you didn't raise beef, you didn't get much. We ate a lot of salt pork that was brought down by train.

"Every family had its own chickens and a milk cow. The cow would give butter and milk and fertilizer for the garden.

The motherless youngster fended for himself while his father was away shrimping. (Courtesy Mary Abadie)

"Grandpa would make wine with raisins and blackberries and we kids would sneak wine and feed the raisins from the cup to the chickens. The chickens would get drunk and grandpa would chase us and pop the bullwhip. We were scared of that whip. All he had to do was pop it.

"Everybody used wood stoves and kerosene lamps. We kids would get two bits [25 cents] a tree to use a rowboat to get oak and ash logs floating down the Mississippi River and tow them to the bank. That is the only trees that they wanted—oak and ash.

"Old man Dupre would pay us kids a penny a bushel for empty oyster shells to put around his camp. We used oars in a pulling skiff to get the oyster shells."

"The back side was all marsh. [People then and still now live on a thin strip of relatively high ground paralleling each side of the Mississippi River. The side facing the river is the "front" and the side facing the marshlands is the "back."] I trapped there as a kid and bought my own school clothes. We caught muskrats and a few mink.

"Rat money was used to build boats and houses. All men would trap in the winter and shrimp in the summer. There were very few oystermen then. Most Croatians came later, after World War II."

Almost all men in coastal marsh communities, young and old, trapped muskrats in the 1930s and 1940s. (Courtesy Jefferson Parish Yearly Review, 1939)

When Brother was old enough, he began working as a deck hand on his uncle's shrimp boat during the summer. He earned 25 cents for every 210-pound barrel of shrimp the boat sold. Shrimpers carried no ice for their catch. Large freight boats, which carried ice, would be anchored on the shrimp grounds. Fishermen would transfer their shrimp to the freight boat, on which they were iced.

Fishermen were not paid by the pound, but rather by volume—a barrel being the standard measurement. The fishermen would receive a ticket for the amount of his catch and the freight boat would haul the shrimp to canneries. Settlement would be made at the end of the month. Fishermen would stay on the grounds for a couple of weeks before coming in, explained Brother.

He has many memories of his early days on shrimp boats. "When I was about nine, I was shrimping with Dad near Battledore Reef in Breton Sound. I was using a bucket on a rope to scoop water to wash the deck on the way in. I lost my balance and fell overboard. The boat kept going. Nets obscured Dad's view and the engine was loud—no muffler—pop, pop, pop.

"It's a funny feeling when you see the boat going out of sight and not turning back. I floated and watched for other boats. Motto Farac was the next boat. It was zigzagging in the rough seas. I swam one way, then the other.

Compared to the modern fishery, shrimping then was a primitive process. (Courtesy Gerald Adkins)

"He didn't slow down. He grabbed me on the run by the wrist. The next day the wrist swole up. We caught up with Dad at the locks because he had to wait until three boats got there. The crews had to help open the locks with a manual capstan. Motto said, 'Joe Pecker; what you doing leaving your son out in the Gulf?' Dad turned red."

He has other memories. In the 1930s, shrimpers struck for a higher price. When some went out shrimping anyway, those on strike poured kerosene on their catches, ruining them.

When Brother graduated the seventh grade from Buras High School, he joined his father on the boat. He worked there until 1945, when he was drafted into the army and spent a year in Saipan. After discharge in 1947, Brother bought his first boat with his uncle, who was a shrimp buyer. "He took 25 percent right off the top of the catch until the boat was paid for."

The boat was a fifty-footer named *Jan*, powered by a 165-horsepower Detroit Diesel, first equipped with a single sixty-five-foot and then later a seventy-five-foot trawl. The engine was much more powerful, he explained, than the 30-horsepower Fairbanks Morse and 60-horsepower Superior engines of his youth.

But the trawls were still made of cotton. Fragile, in spite of being tarred to protect them from sun damage, they tore easily. When he stored them, he dosed them with black pepper in an attempt to deter rats from chewing them up.

Before long, he found out that his uncle was cheating him, so he borrowed enough money from his oldest brother to pay the uncle off. He found someone else to sell to as well: Antoine Alario, who he remembers as "a good man—fair."

Some shrimpers used a drag net, but Brother didn't feel that he needed any sort of try net. Instead he looked for muddy water to find where shrimp were feeding. "My first sale to Antoine was forty-five

Brother grew into a self-confident young man who followed a life on the water without question. (Courtesy Alana Jo Beaugez)

barrels. In one fifteen-minute drag, I saved twenty-five barrels of 10-15 count white shrimp and lost a lot of the drag. We got $45 a barrel."

By 1955, his life was stable enough for him to marry a belle from Braithewaite, Nora Maltese, a union that has produced three daughters, six grandchildren, and six great-grandchildren.

Then in 1962, the direction of his fishing life changed. The year was the worst on record for shrimpers. Brother had to go to work at Empire Machine Works as a mechanic that winter, which he ruefully admits was his only "land job" ever. The next summer he went back to shrimping but added trammel netting for fish to his winter line-up.

At first he used a pulling skiff to fish his trammel nets, targeting redfish, speckled trout, and occasionally black drum. He would string his fish catch up on split palmetto fronds, and that was how they were hauled to the city on trucks.

At that time, only four people in Plaquemines Parish were trammel netters. Fishing for finfish was in its infancy. Three of them, Richard Lilliman, Ruben Bowers, and Jack Rigaud, helped him learn the art of trammel netting. Fish prices were modest. In the 1960s, they received ten cents a pound for small speckled trout, twenty cents for mediums, and thirty cents for large fish, those over two pounds.

Through them, he learned where to fish, times to fish, how to smell fish, and how to look for fish signs—their back-swell in shallow water, muds, and splashing. He started liking netting for fish better than shrimping. More work was involved, but expenses were less and he cleared more money.

Pulling skiffs were sharp-nosed boats propelled by oars by the rower who faced the stern while "pulling." (Courtesy William Knipmeyer)

And he got better at catching fish, some of which he credits to tutoring by Gerald Nix, a transplant from Florida. Fishing gear also improved. First came nylon nets, then monofilament. Finally, gill nets replaced trammel nets.

By the 1970s, finfish was more economically important to Brother than shrimping. Shrimp prices declined and fuel prices rose, and shrimping became his part-time occupation. Brother rapidly became a pioneer in the Louisiana finfishery.

One of Brother's earliest experiments came in 1964, when he began trawling for croakers with a specially designed net. The croakers went to Dave Cerar, an Illinois mink farmer, who used them to feed his animals. He made two or three trips per week, averaging one-half cent per pound for his twenty-six-ton catches.

When Cerar closed the small plant in Empire that processed Brother's catch, another Illinois mink farmer named Ross Wells opened an even larger one. Before selling his catch to Wells, Brother would sort out the large croakers, which he sold for human consumption, as well as any shrimp bycatch. In 1969, Hurricane Camille destroyed the plant.

Brother claims a major role in founding Louisiana's gill net fishery for pompano, the highest priced food fish in the state. Prior to his work, pompano consumed in New Orleans restaurants were imported from other states. He had no predecessors, learning how to fish for them by trial and error or "learning as you go," as he puts it.

Part of the learning curve involved selling as well as catching. In the early years, he peddled his fish directly to fine New Orleans restaurants such as Antoine's, Broussard's, and Brennan's, rather than sell to dealers.

Brother Stipelcovich is probably best known as one of the fathers of the modern pompano fishery.

Another notch in his gun was the development of the mullet roe fishery. In about 1977, he used purse seines with Gene Raffield and Gerry Melvin, both from Florida, to harvest mullet. In 1978, after he got the roe market established, he worked to get the use of purse seines for mullet abolished and helped set up rules for the modern mullet gill net fishery. The first year of the fishery, they received $2.50 to $2.60 per pound—big money in those days.

Brother also claims a role in developing the seine fisheries for redfish and black drum, using small airplanes as spotter craft to locate schools of the fish.

The declaration of redfish as game fish and the legislated loss of the use of gill nets deeply affected Brother, both emotionally and financially. Then, the assault of Hurricane Katrina on lower Plaquemines Parish almost permanently sidelined him. His beloved boat, the *Lady Nora,* washed up on land and his home was destroyed. With the home, Brother lost all his photographs, wood carvings, and tools.

He spent almost a year with a daughter in Childersburg, Alabama. But he finally returned to Empire—his home. There, working out of his elevated house, he still shrimps and fishes, carves miniature models of fishing boats, and rebuilds real, working fishing craft for others.

Nite Owl Special

Cooking in fishing boat galleys is an art form all to its own. Galleys are normally cramped, with small stovetops and few electric kitchen gadgets. If you don't have something you need, you can't run out to the store to get it; you substitute or do without. The cook always has other duties besides cooking, either on the back deck or in the wheelhouse. To top it off, he is cooking in a boat bouncing with the seas. Bowls slide everywhere and liquids slosh out of pots and pans. In spite of this, some amazing dishes are prepared on the water.

Chuck Patrick cooks on the *Nite Owl*, a commercial snapper boat out of Golden Meadow. He is also the vessel's first mate and works on the deck for twelve to sixteen hours a day. Originally hailing from the state of Washington, he cooks in a Cajun/Pacific Northwest fusion style. He modestly calls this dish "fish stuff." I call it poaching. I call it good, too. Chuck whipped up some marvelous meals during the ten days that I spent on the boat with him.

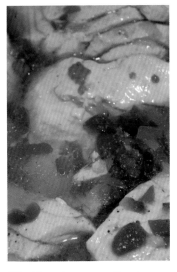

Tip: Beer is an excellent substitute for water in this dish.

Tip: The cooked fish is fragile so handle it with care when turning or removing it from the cooking liquid.

2 lb. fish fillets, at least 1-inch but preferably 2-inches thick
¾ cup water
2 tbsp. olive oil
1 medium onion, diced
1 medium red bell pepper, diced
1 medium green bell pepper, diced
Salt and pepper to taste
1 1-inch-thick slab Velveeta cheese
1 8-oz. block cream cheese

Cut fish into several-inch chunks and set aside. Add water and olive oil to a large saucepan. Add diced onion and red and green bell peppers to the liquid. Cook over a medium heat until the vegetables are soft. Season the liquid to taste with salt and pepper. Add the fish. The fish should be at least half immersed in liquid. If not, transfer to a smaller pan. Bring the liquid to a boil and simmer gently for 5 minutes. Turn the fish carefully and simmer 5 more minutes or until the fish is flakey. Remove the fish from the cooking liquid gently and set it aside in a warm place. Add Velveeta and cream cheese. Stir while cheese is softening to blend thoroughly. The sauce may be spooned over the fish before serving or served at the table. Serves 4.

Bream Fillets Thermidor

When we got married forty-four years ago, both of us came from families that did very little with fish but fry them. Glenda, with her Cajun background, had an occasional courtbouillon, and I didn't even get that as a meat-and-potatoes German. This was my first attempt at something besides frying. The basics of the recipe were something that I fished out of a 1956 Florida Department of Agriculture cookbook, *Florida Seafood Cookery.* The booklet, which we still have, is pretty cool; it has sea turtle recipes in it, and in true Florida fashion, every crawfish recipe in the book refers to spiny lobster. The original recipe didn't specify bream; it just said "fillets." But bream was what I had on hand and they really worked. The tiny fillets rolled up so beautifully. This recipe opened a whole new world of fish cookery for us.

Tip: If another species of fish with larger fillets is used, split the fillets lengthwise so that the rolls are not too wide to sit nested in the milk.

3 lb. bream fillets
2¼ cups milk
1½ tsp. salt
Dash pepper
½ lb. cheddar cheese
½ cup butter
½ cup flour
½ cup lemon juice
1 tbsp. Worcestershire sauce
Paprika

Preheat oven to 350 degrees. Roll each fillet, secure with a toothpick, and place on end in a shallow 2-qt. casserole dish. Pour milk over the fish rolls. Sprinkle them with salt and pepper. Bake about 30 minutes or until the fish flakes easily when tested with a fork but is still moist. While fish is baking, grate cheese coarsely and melt butter. Remove casserole from oven and turn the oven to broil. Drain the milk from the fish and save. Stir the flour into the butter, then slowly stir in the milk. Cook over low heat, stirring constantly until sauce is thickened. Add cheese and stir until melted. Add lemon juice and Worcestershire sauce. Pour the sauce over the baked fillets and sprinkle with paprika. Brown the fish lightly under a broiler. Serves 6-8.

Fantastic Fish

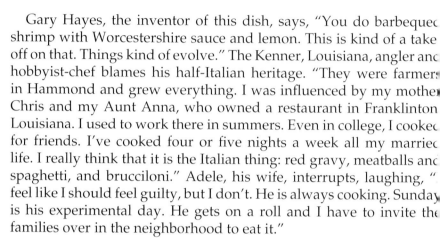

Gary Hayes, the inventor of this dish, says, "You do barbequed shrimp with Worcestershire sauce and lemon. This is kind of a take off on that. Things kind of evolve." The Kenner, Louisiana, angler and hobbyist-chef blames his half-Italian heritage. "They were farmers in Hammond and grew everything. I was influenced by my mother, Chris and my Aunt Anna, who owned a restaurant in Franklinton, Louisiana. I used to work there in summers. Even in college, I cooked for friends. I've cooked four or five nights a week all my married life. I really think that it is the Italian thing: red gravy, meatballs and spaghetti, and brucciloni." Adele, his wife, interrupts, laughing, "I feel like I should feel guilty, but I don't. He is always cooking. Sunday is his experimental day. He gets on a roll and I have to invite the families over in the neighborhood to eat it."

Tip: Make a double batch of this mild, but not under-seasoned, sauce to allow lots for sopping up with bread.

1½ lb. fish fillets
6 tbsp. butter
2 cloves garlic
¼ cup thinly sliced onion
¼ cup thinly sliced bell pepper
1½ tbsp. Worcestershire sauce
1 tsp. salt
1 tsp. black pepper
½ lemon, juiced
⅓ cup white wine
½ cup breadcrumbs
3 green onion tops, chopped

Wash the fillets and set aside. Melt the butter in a small frying pan. Add garlic, onion, bell pepper, Worcestershire sauce, salt, pepper, and lemon juice. Simmer on low until vegetables are soft. Pour into an ovenproof baking dish. Spread the fillets on top of the sauce. Bake at 400 degrees for 20 minutes. Add wine and sprinkle with breadcrumbs. Top with green onions, turn oven to broil, and cook until fish are brown. Check often to prevent burning. Serve with French bread to sop up the sauce. Serves 4.

Curried Fish Roll-Ups

Fish roll-ups can be stuffed with any number of different things, including pesto, crabmeat, breadcrumbs and spinach. And, of course, lots of things can be used to make roll-ups besides fish—chicken, turkey, tortillas and beef. But we like what seems to be everyone else's favorite stuffing—broccoli. It is mild and doesn't overpower the delicate taste of fresh fish. Roll-ups are an easy way to handle fish fillets, without breaking them. This recipe has been in our stable since 1997, when I first published it in the newsletter *Lagniappe*.

2 lb. fish fillets
Salt
½ tsp. white pepper
1 6-oz. package mozzarella cheese slices
2 10-oz. boxes frozen broccoli spears, cooked and drained
1 cup boiling water
1 chicken bouillon cube
½ cup dry white wine
2 cloves garlic, minced
½ tsp. Creole seasoning
1 tbsp. cornstarch
1 ½ tsp. curry powder
2 tbsp. water

Spread fillets on wax paper and sprinkle with salt and white pepper. Cut mozzarella cheese slices in thirds lengthwise. Place a strip of cheese and equal amounts of broccoli spears on each fillet. Roll up the fillets and secure them with toothpicks. Pour boiling water in a 10-inch saucepan or deep-sided skillet and dissolve the bouillon cube. Add wine, garlic, and Creole seasoning. Add the roll-ups to the liquid and bring it back to a boil. Cover and cook over a medium heat for 10-13 minutes or until the fish flakes easily with a fork. Carefully remove roll-ups to a hot platter. Combine cornstarch, curry powder, and 2 tbsp. water. Mix well and add to the remaining liquid in the pan. Cook until thick and smooth, stirring constantly. Pour sauce over roll-ups. Serves 4-6.

Blaff

This is an interesting dish. It hails from the French West Indies, specifically Martinique, where it is very popular. This version is for head-on fish, although it can also be made with skin-on steaks. A variety of fish is used in its homeland—everything from snappers and groupers to marlin and swordfish, even haddock or cod, which are surprisingly often imported to the islands. Essentially, it is poached fish that have been marinated in a lime juice solution. Hot peppers, the hotter the better, are necessary, although the final dish is more aromatically seasoned than spicy. The name supposedly comes from the sound that the fish makes when it hits the boiling liquid. I think it goes "pfsooph" rather than "blaff." You be the judge.

2 17-18-inch snappers
2 tbsp. salt
2 tsp. pepper
2 tsp. ground allspice, divided
3 cloves garlic, minced
2 habanero peppers, quartered
6 limes, juiced and divided
1 gal. water
1 small onion, diced
1 sprig fresh thyme, chopped
2 green onions, chopped
1 sprig parsley, chopped

Remove the guts and gills of the fish and scale them. Leave the heads on. Set aside. Prepare a marinade of salt, pepper, 1 tsp. allspice, 1 clove garlic, habanero peppers, and juice of 3 limes. Place the fish in a bowl, cover with the marinade, and refrigerate for 1 hour. When ready to cook, remove the fish from the bowl, place the marinade and all of the ingredients, except the remainder of the lime juice and the fish in a heavy pot and bring to a boil. When the water is at a rolling boil, place the fish in the liquid (listen for the blaff!) and allow it to return to a boil. Remove fish when done, approximately 10 minutes, and serve with cooking liquid and remaining lime juice. Serve over cooked rice. This dish cooks quicker than you think. Watch it carefully. Serves 4.

The road's end in the marsh—Delacroix, Louisiana, before Hurricane Katrina.

Index